INTEGRATED REVIEW WORKSHEETS

DOUGLAS R. NERING
Ivy Tech Community College

to accompany

MYMATHLAB FOR QUANTITATIVE REASONING WITH INTEGRATED REVIEW

Jeffery O. Bennett
University of Colorado at Boulder

William Briggs
University of Colorado at Denver

PEARSON

Boston   Columbus   Indianapolis   New York   San Francisco   Upper Saddle River
Amsterdam   Cape Town   Dubai   London   Madrid   Milan   Munich   Paris   Montreal   Toronto
Delhi   Mexico City   São Paulo   Sydney   Hong Kong   Seoul   Singapore   Taipei   Tokyo

ISBN-13: 978-0-321-98785-3
ISBN-10: 0-321-98785-3

1 2 3 4 5 6 EBM 17 16 15 14

www.pearsonhighered.com

PEARSON

CONTENTS

Integrated Review Worksheets

Using and Understanding Mathematics with Integrated Review

Name:_____ Date:_____

Instructor:_____ Section:_____

Chapter 1 Thinking Critically

Learning Objectives
Objective 1.R.1 – Compare real numbers Objective 1.R.2 – Evaluate an algebraic expression Objective 1.R.3 – Classify real numbers

Objective 1.R.1 – Compare real numbers

In mathematics, we study statements - sentences that are either true or false but not both. For example, 8 is an even integer (true statement) and 10 is an odd integer (false statement).

The purpose of this section is to use integer relationships and inequalities to advance our logical reasoning skills. Our understanding of these integer relationships combined with inequality interpretation will help us recognize logical arguments that are either well supported or poorly supported by facts or assumptions (integer facts). To help analyze and interpret inequalities, we need to know the meaning of the following inequality symbols:

> **> indicates greater than**
> **< indicates less than**
> **≥ indicates greater than or equal to**
> **≤ indicates less than or equal to**

Guided Examples   Practice

Guided Examples	Practice
1) Is the statement $5 \geq 8$ true or false (a fallacy)? The inequality symbol communicates that 5 is greater than or equal to 8. This statement is **false (a fallacy)** since 5 has a unit value less than 8.	1) Is the statement $-3 \leq 5$ true or false (a fallacy)?
	2) Is the statement $6 > 3$ true or false (a fallacy)?
2) Is the statement $-3 < 0$ true or false (a fallacy)? The inequality symbol communicates that -3 is less than 0. This statement is **true** since -3 has a unit value less than 0.	3) Is the statement $12 > 12$ true or false (a fallacy)?
3) Is the statement $15 \geq 15$ true or false (a fallacy)? The inequality symbol communicates that 15 is greater than or equal to 15. This is **true** since 15 has a unit value that is equal to 15.	4) All the Presidents of the United States have been male so men are better qualified than women to be President (true or false – a fallacy).

Objective 1.R.2 – Evaluate an algebraic expression

In this section, we will evaluate an algebraic expression to aid in our understanding and mastery of propositions and logical connectors.

Proposition review

A proposition is a declarative statement that is either true or false (not both). We will explore algebraic concepts to improve our understanding of these logic principles.

For example, $8 + 12 = 20$ is a proposition (a statement) that is true.

Another logic concept is the negation. It transforms a proposition into its opposite truth value (true to false and false to true). For our algebraic example, we know $8 + 12 = 20$. The negation of this example is $8 + 12 \neq 20$.

Guided Examples

Practice

1) Is the statement $12 - 4 = 8$ true or false? This statement is true. 2) Write the negation of $8 + 2 = 10$. A negation makes the opposite claim. The negation of is $8 + 2 \neq 10$.	1) Is the statement $9 + 3 = 11$ true or false? 2) Write the negation of $6 + 4 = 10$.

Logic connector review

Logical connectors are used to join or connect two ideas. These two ideas or statements can be joined together with the connector AND. This creates a compound statement that is **true only when both parts** of the compound statement are true.

The other type of logical connector we will review is OR. This connector also joins two ideas or statements together into a compound statement. This compound statement is **true when either one or both parts** of the compound statement are true.

Guided Examples

Practice

Determine if the compound statement is true or false. 3) $3 + 9 = 12$ AND $3 \cdot 5 = 12$. Is this entire statement true or false? Since we use the connector AND, both parts of the compound statement need to be true for the entire statement to be true. $3 + 9 = 12$ is true and $3 \cdot 5 = 12$ is false, so this entire statement is false.	Determine if the compound statement is true or false. 3) $5 - (-3) = 8$ AND $2 \cdot 8 = 16$ 4) $-6 \cdot -2 = 12$ AND $(4 \cdot 7) - 10 = 14$ 5) Green beans are vegetables AND oranges are fruit.

4) $4+9=15$ OR $-5 \cdot -3 = 15$ Is this entire statement true or false? Even though $4+9=15$ is a false statement, since $-5 \cdot -3 = 15$ is a true statement, so the entire compound statement is true.	6) $3 \cdot 2 = 6$ OR $10+10=0$ 7) $5 \cdot 5 = -25$ OR $7-1=8$ 8) Indiana is a state or California is a country.

Objective 1.R.3 – Classify real numbers

There are a number of different ways to classify numbers. This section will introduce the most common classifications of real numbers so you can explore, apply and master logic concepts. Specifically, we will investigate real number classification as it applies to set relationships and set theory, as well as, Venn diagram design and construction.

Natural Numbers: The natural numbers start with one and compose the numbers $1, 2, 3, 4...$ and so on. Zero is not considered a "natural number."

Whole Numbers: The whole numbers start with zero and compose the numbers $0, 1, 2, 3, 4, ...$ and so on (the natural numbers including zero).

Integers: The integers are the whole numbers and their opposites (the positive whole numbers, the negative whole numbers, including zero). Integers contain $...-3, -2, -1, 0, 1, 2, 3...$

Rational Numbers: The rational numbers include all the integers, plus fractions that consist of two irrational numbers (that can be written as a quotient of integers with the denominator $\neq 0$), or terminating decimals and repeating decimals. Rational numbers contain $...-3, -2.5, -2, -1, 0, \frac{1}{3}, \frac{1}{2}, 1, 2, 3...$

Irrational Numbers: An irrational number is a number with a decimal that neither terminates nor repeats. For example, the $\sqrt{2}$ is a decimal that does not repeat itself, but that continues infinitely. The value π (pi) is classified as an irrational number as well.

Real Numbers: All the rational numbers and all the irrational numbers together form the real numbers.

This narrative description of real number classification is visually demonstrated using the Venn diagram below.

Guided Examples	Practice
1) Identify the set(s) that describe -5 .	Identify the specified set.
	1) Identify the set(s) that describe $\frac{3}{4}$.
Since -5 is an integer, then it is a member of the set(s): integer, rational, and real.	
	2) Identify the set(s) that describe 7 .
	3) I can buy an irrational amount of apples (True or False).
	4) I can buy an amount of apples that is a natural number (True or False).

Chapter 2 Approaches to Problem Solving

Learning Objectives
Objective 2.R.1 – Multiply and divide fractions Objective 2.R.2 – Write a number in scientific or standard notation

Objective 2.R.1 – Multiply and divide fractions

Dimensional analysis is a mathematical system using conversion factors to move from one unit of measurement to a different unit of measurement. Typically, this involves the multiplication of two or more fractions using a specific dimensional analysis format. Our emphasis is on the precise application of multiplying and dividing fractions to become skilled at dimensional analysis calculations and develop an accurate interpretation of the data.

Multiply fractions review

There are four steps to multiplying two fractions:

- multiply the two numerators (top numbers) together
- multiply the two denominators (bottom numbers) together.
- The resulting new numerator is the product of the numerator in the first fraction and the numerator in the second fraction. The resulting new denominator is the product of the denominator in the first fraction and denominator in the second fraction.
- If need be, simplify the resulting fraction.

Guided Examples	Practice

Multiply the fractions:	Multiply the fractions
1) $\dfrac{6}{1} \cdot \dfrac{3}{5} =$ $\dfrac{6 \cdot 3}{1 \cdot 5} = \dfrac{18}{5}$ Multiply the numerators (top numbers), then multiply the denominators (bottom numbers). The resulting fraction is your answer.	1) $\dfrac{2}{3} \cdot \dfrac{5}{7} =$
2) $\dfrac{3}{7} \cdot \dfrac{2}{5} =$ $\dfrac{3 \cdot 2}{7 \cdot 5} = \dfrac{6}{35}$ Multiply the numerators (top numbers), then multiply the denominators (bottom numbers). The resulting fraction is your answer.	2) $\dfrac{3}{1} \cdot \dfrac{2}{5} =$

Divide fractions review

There are three steps to dividing two fractions:

- Modify the second fraction first by turning it upside-down (the reciprocal)
- Change the operation from division to multiplication then multiply the first fraction by the reciprocal (multiply across – numerator to numerator and denominator to denominator)
- If need be, simplify the resulting fraction.

Guided Examples   Practice

Divide the fractions:	Divide the fractions:
3) $\dfrac{4}{5} \div \dfrac{7}{3} = \dfrac{4}{5} \cdot \dfrac{3}{7} = \dfrac{4 \cdot 3}{5 \cdot 7} = \dfrac{12}{35}$ You must modify the second fraction. $\dfrac{7}{3}$ becomes $\dfrac{3}{7}$. Next, change the operation from division to multiplication. Multiply the first fraction and the modified second fraction Multiply across the numerators (top numbers) and the denominators (bottom numbers).	3) $\dfrac{2}{7} \div \dfrac{1}{5} =$
4) $\dfrac{2}{11} \div \dfrac{5}{3} = \dfrac{2}{11} \cdot \dfrac{3}{5} = \dfrac{2 \cdot 3}{11 \cdot 5} = \dfrac{6}{55}$ You must modify the second fraction. $\dfrac{5}{3}$ becomes $\dfrac{3}{5}$. Next, change the operation from division to multiplication. Multiply the first fraction and the modified second fraction Multiply across the numerators (top numbers) and the denominators (bottom numbers).	4) $\dfrac{4}{1} \div \dfrac{3}{5} =$

Objective 2.R.2 – Write a number in scientific or standard notation

Scientific Notation was developed in order to easily represent numbers that are either very large or very small. To simplify matters when describing, writing or calculating with very large or very small numbers, we often use scientific notation to represent these numerical extremes.

This learning objective will focus on converting between scientific and standard notation in order to more accurately solve dimensional analysis challenges.

Converting from standard notation to scientific notation review

A number is in scientific notation when it is broken up as the product of two parts. The general form of scientific notation is $a \times 10^n$ where a is greater than or equal to 1 and less than 10, and n is an integer. For example, the value 4.7×10^3 is written in scientific notation where 4.7 is the coefficient and 10^3 is the power of ten.

There are three elements to consider when converting a number from standard notation to scientific notation:
- Place the decimal point such that there is **one nonzero digit to the left of the decimal point**.
- Count the number of decimal places the decimal has "moved" from the original number. This will determine the absolute value of the exponent of the 10.
- If the original number is less than 1, the exponent is negative; if the original number is greater than 1, the exponent is positive.

Guided Examples   Practice

Convert from standard notation to scientific notation	Convert from standard notation to scientific notation	
1) $2,350,000 = 2.35 \times 10^6$ Since the decimal point is placed after the zeros, it is move to between the 2 and 3, you get a coefficient of 2.35. The decimal point has moved 6 decimal places so the power of 10 is 10^6. The original number is greater than 1, so the exponent is positive. 2) $2116 = 2.116 \times 10^3$ 3) $0.00082 = 8.2 \times 10^{-4}$ Since the decimal point is before the zeros, it is moved to between the 8 and 2, you get a coefficient of 8.2. The decimal point has moved 4 decimal places so the power of 10 is 10^4. The original number is less than 1, so the exponent is negative. 4) $0.0019 = 1.9 \times 10^{-3}$	1) $328,000 =$ 2) $14,629 =$ 3) $0.0034 =$ 4) $0.0000058 =$	

Converting from scientific notation to standard notation review

There are two elements to consider when converting a number from scientific notation to standard notation:
- Move the decimal point to the right for positive exponents of 10. The exponent tells you how many places to move. The positive exponent indicates the resulting number (answer) is a large number (greater than 1).
- Move the decimal point to the left for negative exponents of 10. The exponent tells you how many places to move. The negative exponent indicates the resulting number (answer) is a small number (less than 1).

Guided Examples   Practice

Convert from scientific notation to standard form	Convert from scientific notation to standard form
5) $7.21 \times 10^4 = 72,100$ Because the exponent is positive, move the decimal point to the right 4 (the exponent value) places. The positive exponent indicates the answer is a large number (greater than 1). 6) $4.358 \times 10^{-3} = 0.004358$ Because the exponent is negative, move the decimal point to the left 3 (the exponent value) places. The negative exponent indicates the answer is a small number (less than 1).	5. $6.82 \cdot 10^3 =$ 6) $7.1 \cdot 10^5 =$ 7) $9.11 \cdot 10^{-4} =$ 8) $2.185 \cdot 10^{-2} =$

Name:_____Date:_____

Instructor:_____ Section:_____

Chapter 3 Numbers in the Real World

Learning Objectives
Objective 3.R.1 – Simplify fractions Objective 3.R.2 – Multiply and divide by decimal values to introduce percentages Objective 3.R.3 – Percent conversions Objective 3.R.4 – Multiply or divide using scientific notation Objective 3.R.5 – Decimal rounding

Objective 3.R.1 – Simplify fractions

In this chapter, your challenge will be to determine, explore and interpret ratios. A ratio shows the relative size of two (or more) values. Ratios can be communicated in different forms using the example as follows:

Your survey your class and 7 of your colleagues wear glasses out of the 20 people in your class. Determine and write the ratio of people who wear glasses to the total class. Your answer can be written in several forms as follows:

- $7 : 20$ (7 people wear glasses in your class of 20)

- $\dfrac{7}{20}$ (7 people wear glasses in your class of 20)

- 0.35 or $35\% (7 \div 20)$ of the people in your class wear glasses

At times, a ratio will need to be reduced to its simplest form using division. For example, you survey the 20 people in your class. It turns out that 8 of your colleagues are over the age of 25. Determine and write the ratio of people over the age of 25 to the total people in your class. Your answer can be written as $8 : 20$ or $\dfrac{8}{20}$ or $40\% (8 \div 20)$.

Because the ratio written in fraction form is not written in simplest form, we need to simply the ratio. This is a two-step process that requires you to do the following:

- Find the Greatest Common Factor (GCF) of the numerator and denominator
- Divide the numerator and the denominator by the GCF

In this case, 4 is the largest number (GCF) common to both 8 and 20. We can now determine the ratio in simplest form $\dfrac{8}{20} = \dfrac{8 \div 4}{20 \div 4} = \dfrac{2}{5}$.

Guided Examples   Practice

Reduce the ratio	Reduce the ratio
1) $\dfrac{24}{32}$	1) $\dfrac{15}{55} =$
$\dfrac{24}{32} = \dfrac{24 \div 8}{32 \div 8} = \dfrac{3}{4}$ Identify the Greatest Common Factor (GCF) common to both the numerator and denominator. The GCF is 8. Next, divide both the numerator and denominator by the GCF to determine the ratio in simplest form.	2) $\dfrac{2}{10} =$

2) $\dfrac{6}{15}$

Identify the Greatest Common Factor (GCF) common to both the numerator and denominator. The GCF is 3. Next, divide both the numerator and denominator by the GCF to determine the ratio in simplest form.

$$\frac{6}{15} = \frac{6 \div 3}{15 \div 3} = \frac{2}{5}$$

3) You survey your class and 4 like vanilla ice cream and 14 like chocolate ice cream. Write the ratio of people who like vanilla to people who like chocolate ice cream, then write the ratio in simplest form.

Objective 3.R.2 – Multiply and divide by decimal values to introduce percentages

In this section, you will apply percentages to real world examples. This type of math modeling explores relative change and absolute change. In order to accurately calculate and interpret your results, we should briefly discuss percentages in general, then four conversion ideas.

The word "percent" means "per 100" or "out of 100" or "divided by 100". For example, 39% means 39 out of 100. Percent's can also be expressed as fractions or decimals, since they are often used to indicate some part of a whole. So, 39% can also be written as $\frac{39}{100}$ or .39.

There are four conversion types you should be familiar with. They are as follows:

- To convert a percentage to a fraction: The 100 becomes the denominator and do not use the % symbol.
  - Example: $23\% = \frac{23}{100}$ (NOTE: If necessary, reduce the fraction to simplest terms.)
- To convert a percentage to a decimal: Move the decimal point two places to the left and do not use the % symbol.
  - Example: $75\% = 0.75$
- To convert a decimal to a percentage: Move the decimal point to places to the right and use the % symbol.
  - Example: $0.81 = 81\%$
- To convert a fraction to a percentage: Divide the numerator by the denominator to change the fraction to decimal form. Then, move the decimal point two places to the right and use the % symbol.
  - Example: $\frac{4}{16} = 4 \div 16 = 0.25 = 25\%$

Guided Examples	Practice
Convert and calculate the following examples:	Convert and calculate the following examples:
1) Convert 19% to a fraction. 100 is the denominator and do not use the % symbol, so the conversion is $\frac{19}{100}$.	1) Convert 83% to a fraction.
2) Convert 46% to a decimal. Move the decimal point two places to the left and do not use the % symbol, so the conversion is 0.46	2) Convert 14% to a decimal.
3) Convert 0.028 to a percentage. Move the decimal point two places to the right and use the % symbol, so the conversion is 28%	3) Convert 0.57 to a percentage.
4) Convert $\frac{7}{8}$ to a percentage. Divide the numerator by the denominator. The result is $7 \div 8 = 0.875$. Now, move the decimal point two places to the right and use the % symbol, so the final conversion is 87.5%.	4) Convert $\frac{3}{5}$ to a percentage.
	5) You decide to purchase an item with an original cost of $250 that is discounted 30%. What is the dollar amount of the discount?

5) You make a purchase of $48 and are required to pay 6% sales tax. What is the dollar amount of the sales tax?

Convert the 6% to a decimal and multiply the two values. The sales tax calculation is $48 \cdot 0.06 = \$2.88$.

6) You make an online purchase of $120 and must pay 8% shipping and handling fees. What is the dollar amount of the S & H fees?

Convert 8% to a decimal and multiply the two values. The S & H calculation is $120 \cdot 0.08 = \$9.60$.

7) Your dinner costs $36 and you want to leave a 20% tip. What is the dollar amount of the tip?

Convert 20% to a decimal and multiply the two values. The tip calculation is $36 \cdot 0.20 = \$7.20$.

6) Your weekly earnings of $600 is taxed at the rate of 15%. What is the dollar amount of your taxes?

7) Because of the concern of global warming, everyone is asked to reduce their carbon footprint. An average U.S. person has a carbon footprint of 20 tons. If you are asked to reduce your carbon footprint by 10%, what is the amount of reduction in tons?

Objective 3.R.3 – Percent conversions

In order to analyze the uses and abuses of percentages, you will develop the ability to accurately convert from percent to decimal and decimals to percent. Additionally, your precise understanding of percent conversions will aid in your ability to accurately reason and logic numerical challenges.

Converting from percent to decimal requires division. The percent symbol means "divided by 100". For example, 15% is $\dfrac{15}{100} = 0.15$. The short way to convert from a percent to a decimal is to move the decimal point two places to the left and remove the percent symbol.

Converting from decimal to percent requires multiplication by 100. For example, $0.825 \cdot 100 = 82.5\%$. The short way to convert from decimal to percent is to move the decimal point two places to the right and add the percent symbol.

Guided Examples	Practice
Convert the following from percent to decimal: 1) $8.5\% = 0.085$ $8.5\% = \dfrac{8.5}{100} = 0.085$ or move the decimal point two places to the left and remove the percent symbol. 2) $200\% = 2.0$ or 2 No decimal point is given, so it is placed at the end of the number (after the 0). The decimal point is moved two places to the left and remove the percent symbol. Convert the following from decimal to percent: 3) $0.068 = 0.068 \cdot 100 = 6.8\%$ Multiply the decimal by 100 or move the decimal point two places to the right, and add the percent symbol.	Convert the following from percent to decimal: 1) $7.38\% =$ 2) $75\% =$ 3) The price of gas today is 200% of the price last year. Does that mean the price has doubled? Convert the following from decimal to percent: 4) $0.25 =$ 5) $0.05 =$ 6) Can it be true that a 10 year old child weighs 30% more than a 6 year old child?

Objective 3.R.4 – Multiply or divide using scientific notation format

The scientific notation format was developed to easily represent very large or very small numbers. This section reviews the conversion of writing very large or very small numbers (standard form) in scientific notation form, especially as it relates to federal spending and the federal deficit.

A number in scientific notation is written as the product of a number (integer or decimal) and a power of 10 in the form $a \times 10^b$ where (for us) a is a rational number greater than 1 and less than 10 and b (the exponent) is an integer. There are three parts to consider when converting from standard form (decimal) to scientific notation:

- the rational number a
- the value of the exponent
- is the exponent positive or negative

The rational number (a) in scientific notation form always has only one non-zero digit to the left of the decimal point. The absolute value of the exponent indicates how many "spaces" you moved your decimal point.

Additionally, if the original number you are working with is greater than 1, the exponent is positive and if the original number you are working with is less than 1, the exponent is negative.

Guided Examples	Practice
Write the following numbers in scientific notation form: 1) $4,200,000 = 4.2 \times 10^6$ Move the decimal point so there is only one digit to the left of the decimal point (4.2), you moved the decimal point 6 places (exponent 6) and since your beginning number is large (exponent is positive). 2) $3,560.2 = 3.5602 \times 10^3$ 3) $0.00081 = 8.1 \times 10^{-4}$ Move the decimal point so there is only one digit to the left of the decimal point (8.1), you moved the decimal point 4 places (absolute value of the exponent is 4) and since your beginning number is small (less than 1 – the exponent is negative).	Write the following numbers in scientific notation form: 1) $315,000 =$ 2) Currently, the federal deficit is approximately $\$17,000,000,000,000$ (17 trillion dollars). Write this number in scientific notation. 3) $0.0037 =$ 4) The diameter of a typical bacterium is 0.000001 meter. Write this number in scientific notation.

To convert from scientific notation to standard form, there is a pattern:

- If the exponent is positive, the exponent designates you move the decimal point to the right by the exponent value.
- If the exponent is negative, the exponent designates you move the decimal point to the left by the exponent value.

<u>Guided Examples</u>   <u>Practice</u>

Write the following numbers in standard form:	Write the following numbers in standard form:
5) $3.8 \times 10^4 = 38,000$ The exponent is positive, so move the decimal point to the right 4 places. The positive exponent indicates a "large" number. 6) $9.23 \times 10^6 = 9,230,000$ The exponent is positive, so move the decimal point to the right 6 places. 7) $4.71 \times 10^{-3} = 0.00471$ The exponent is negative, so move the decimal point 3 places to the left. The negative exponent indicates a "small" number. 8) $8.9 \times 10^{-6} = 0.0000089$ The exponent is negative, so move the decimal point 6 places to the left.	5) $2.9 \times 10^3 =$ 6) The Earth is 1.496×10^8 km from the Sun. Write this number in standard form. 7) $4.3 \times 10^{-2} =$ 8) An average cell has an approximate diameter of 6×10^{-6} meters. Write this number in standard form.

For our federal revenue or deficit calculations involving the population of the United States may require us to combine (multiplying or dividing) large numbers.

If we are multiplying two values in scientific notation form,
- Multiply the rational numbers
- Add the exponents

If we are dividing two values in scientific notation form,
- Divide the rational numbers
- Subtract the exponents

Guided Examples   Practice

Multiply the scientific notation values:	Multiply the scientific notation values:
9) $(3.1\times10^5)(2.5\times10^3) = 7.75\times10^8$ Multiply the values: $3.1\cdot2.5 = 7.75$ Add the exponents: $10^{5+3} = 10^8$ 10) $(4.1\times10^5)(2\times10^7) = 8.2\times10^{12}$ Multiply the values: $4.1\cdot2 = 8.2$ Add the exponents: $10^{5+7} = 10^{12}$	9) $(3.2\times10^6)(2.5\times10^2) =$ 10) $(1.6\times10^3)(6\times10^7) =$
Divide the scientific notation values: 11) $\dfrac{9.3\times10^8}{3.1\times10^2} = 3\times10^6$ Divide the values: $9.3\div3.1 = 3$ Subtract the exponents: $10^{8-2} = 10^6$ 12) $\dfrac{8.64\times10^{14}}{3.2\times10^3} = 2.7\times10^{11}$ Divide the values: $8.64\div3.2 = 2.7$ Subtract the exponents: $10^{14-3} = 10^{11}$	Divide the scientific notation values: 11) $\dfrac{8.6\times10^{12}}{4.3\times10^3} =$ 12) $\dfrac{9.2\times10^5}{5\times10^2} =$

Objective 3.R.5 – Decimal rounding

Understanding the concept of significant digits will better equip you to assess the considerations required in dealing with numerical literacy and calculation challenges. In this regard, there are two areas for us to explore. Place value and rounding are used to describe results with accuracy and precision.

Place value refers to the positioning of either a single digit in a whole number or in a number containing a decimal. The key to determining place value is to become familiar with the below place value map.

Place value chart											
millions	hundred thousands	ten thousands	thousands	hundreds	tens	ones	decimal	tenths	hundredths	thousandths	ten thousandths

Next, we will use the place value chart to develop an understanding of the process of rounding numbers as it applies to the rounding with significant digits.

The process of rounding numbers is a two-step process as follows:

- Step 1: Decide which place value is most important.
- Step 2: Look at the number in the place to the *right*.
  - If the value in this next place is less than five $(0 \text{ to } 4)$, there is no change to the place value.
  - If the value in this next place is greater than or equal to five $(5 \text{ to } 9)$, there is a change of $+1$ to the place value.
  - All values to the right of the designated place value are no longer used in the rounded answer and zeros are used for place holder values.

Guided Examples   Practice

Round the following numbers to the value indicated	Round the following numbers to the value indicated
1) 2643.7 to the nearest hundred is 2600 . The hundred place value is designated. This is the value 6 . Look at the number in the place to the right of the 6 . This is the value 4 . Since 4 is less than 5 , then the value 6 is not changed. The 43.7 is not used in the rounded answer (replaced with zeros).	1) 428.1 to the nearest ten. 2) 526,034 to the nearest ten thousand. 3) 9833.5 to the nearest hundred.

2) 37.923 to the nearest hundredth is 37.92 .

The hundredth place value is designated. This is the value 2 . Look at the number in the place to the right of the 2 . This is the value 3 . Since 3 is less than 5 , then the value 2 is not changed. The 3 is not used in the rounded answer.

3) 14,692.7 to the neared thousand is 15,000 .

The thousand place value is designated. This is the value 4 . Look at the number in the place to the right of the 4 . This is the value 6 . Since 6 is greater than 5 , then the value 4 is changed to 5 . The 692.7 is not used in the rounded answer (replaced with zeros).

4) 8.62 to the nearest tenth is 8.6 .

The tenth place value is designated. This is the value 6 . Look at the number in the place to the right of the 6 . This is the value 2 . Since 2 is less than 5 , then the value 6 is not changed. The 2 is not used in the rounded answer.

4)   8.2137 to the nearest thousandth

5)   439.25 to the nearest tenth

6)   45.338 to the nearest hundredth

Chapter 4 Managing Money

Learning Objectives

Objective 4.R.1 – Use order of operations for real numbers
Objective 4.R.2 – Solve proportions
Objective 4.R.3 – Find the amount, base, or percent in a percent problem by solving an equation
Objective 4.R.4 – Solve linear equations using addition and multiplication
Objective 4.R.5 – Evaluate radical expressions

Objective 4.R.1 – Use order of operations for real numbers

PEMDAS Review

Taking control of your finances requires you to keep track of monthly expenses. To annualize (12 months) these expenses, you multiply by 12. Some expenses are semiannual (twice a year), so you may add expenses then multiply by 2. Adjusting cash flow or a budget may require you to work with negative numbers. This review section explores all these rules so you are able to confidently make personal finance calculations in this section.

You will frequently use the PEMDAS acronym to help guide your personal finance decision making when evaluating algebraic expressions:

Parentheses | Exponents | Multiplication | Division | Addition | Subtraction

1. Perform the operations inside a parentheses first
2. Then exponents
3. Then multiplication and division, from left to right
4. Then addition and subtraction, from left to right

Integer Review

1. **Adding Rules with examples:**

Positive + Positive = Positive: $4 + 7 = 11$
Negative + Negative = Negative: $(-5) + (-2) = -7$

Sum of a negative and a positive number: You need to subtract the absolute values of the two numbers then use the sign of the number with the larger absolute value.

$(-11) + 2 = -9$
$4 + (-5) = -1$
$(-2) + 9 = 7$
$8 + (-5) = 3$

2. **Subtracting Rules with examples:**

Negative - Positive = Negative: $(-7)-6=-7+(-6)=-13$

Positive - Negative = Positive + Positive = Positive: $7-(-2)=7+2=9$

Negative - Negative = Negative + Positive = You need to subtract the absolute values of the two numbers then use the sign of the number with the larger absolute value. *(NOTE: Two negatives next to each other become positive)*
$(-2)-(-8)=(-2)+8=6$
$(-3)-(-7)=(-3)+7=4$

3. **Multiplying Rules with examples:**

Positive × Positive = Positive: $8 \cdot 2 = 16$

Negative × Negative = Positive: $(-3) \cdot (-7) = 21$

Negative × Positive = Negative: $(-8) \cdot 5 = -40$

Positive × Negative = Negative: $5 \cdot (-3) = -15$

4. **Dividing Rules with examples:**

Positive ÷ Positive = Positive: $18 \div 3 = 6$

Negative ÷ Negative = Positive: $(-36) \div (-4) = 9$

Negative ÷ Positive = Negative: $(-24) \div 3 = -8$

Positive ÷ Negative = Negative: $15 \div (-3) = -5$

Guided Examples	Practice
1) Evaluate the expression $5 \cdot (3+6)$ Because of the order of operations, you will perform operations inside the parentheses first, $3+6=9$ Next, you multiply $5 \cdot 9$. The value of the expression is 45. 2) Calculate the total when the following college expenses are paid twice a year: Tuition of 3000, 800 in student fees and 350 for textbooks. The expression can be set up as $2(3,000+800+350)$. Because of the order of operations, you will perform operations inside the parentheses first. So, $2(4,150)$ totals $\$8,300$.	1) $(8+12) \cdot 5$ 2) $5+2(3+6)^2$ 3) You spend $\$20$ a week for coffee and $\$180$ per month for groceries. If a month is four weeks long, calculate your total coffee and grocery expense? 4) You decide to lease/buy a car with the following conditions: a one-time fee of $\$300$, a down payment of $\$2000$ and monthly payments of $\$210$. What is the annual (12 month cost) of this plan?

3) Determine if you have a positive or negative monthly cash flow with the following details:

Income: Part time job $\$400$ per month, scholarship $\$300$ per month and a student loan of $\$250$ per month.

Expenses: Rent $\$725$ per month, groceries $\$65$ per month, phone $\$60$ per month and entertainment is $\$180$ per month.

The expression can be set up as $(400+300+250)-(725+65+60+180)$. This becomes $950-1030=-80$. This represents a negative cash flow of $\$80$ per month.

5) $15+3=$

6) $20-(-5)=$

7) $-9+(-2)=$

8) $-6 \cdot 10=$

9) $30 \div -3=$

10) $9+(-4)=$

11) $-5 \cdot -2=$

12) $-20 \div -2=$

13) $-18+12=$

Objective 4.R.2 – Solve Proportions

A proportion is an equation which states that two ratios (fractions) are equal to each other. If one term of a proportion is not known (designated by the variable x), the concept of cross multiplication can be used to find the value of the unknown term.

The numerator is the top number in a fraction and the denominator is the bottom number in a fraction. For example, in the fraction $\frac{3}{8}$, 3 is the numerator and 8 is the denominator.

To cross multiply, take the first fraction's numerator and multiply it by the second fraction's denominator. Then take the first fraction's denominator and multiply it by the second fraction's numerator.

When the terms of a proportion are cross multiplied, the terms are set these equal to each other and the fraction form is no longer used.

Guided Examples	Practice
1) Solve the proportion for : $\dfrac{4}{12} = \dfrac{9}{x}$ To set up the cross multiplication, $4 \cdot x = 12 \cdot 9$. Multiply the terms on each side becomes $4x = 108$. To isolate the variable and solve for x, divide both sides of the equation by 4 and you get $x = 27$. 2) Solve the proportion for x: $\dfrac{3}{8} = \dfrac{6}{x+4}$ To set up the cross multiplication, make sure the 3 is distributed to all terms in the denominator, so $3(x+4) = 8 \cdot 6$. Next, distribute the 3 to all terms in the parentheses by multiplication. When you distribute, the parentheses are no longer required. You get $3x + 12 = 48$ To isolate the variable, you do the opposite operation of add 12, which is to subtract 12 from both sides of the equation. The result is $3x = 36$. Finally, you do the opposite operation of multiply by 3, which is to divide both sides by 3. The result is $x = 12$.	1) Solve the proportion $\dfrac{9}{36} = \dfrac{4}{x}$ 2) Solve the proportion $\dfrac{15}{x} = \dfrac{20}{12}$ 3) Solve the proportion $\dfrac{2x+3}{6} = \dfrac{5}{2}$ HINT: Make sure you multiply the 2 to both terms in the parentheses $2(2x+3)$. 4) Set up a proportion to solve. There are 12 students who wear glasses in your class of 20. If you go to a class with 50 students, how many will be wearing glasses? HINT: Be sure to "line up" the units in both your ratios: $\dfrac{\text{students wear glasses}}{\text{students in class}} = \dfrac{\text{students wear glasses}}{\text{students in class}}$

Name:_____Date:_____

Instructor:_____Section:_____

Objective 4.R.3 – *Find the amount, base, or percent in a percent problem by solving an equation*

Percentages refer to fractions of a whole. In our application of earning a percentage of interest on money invested, the percentage refers to earning (or receiving) a fraction of the money you invested in the form of interest earned the bank will pay to you.

To multiply a value by a percent requires that the percent be converted to a decimal so you can multiply the values. For example, to complete the operation $2.6\% \cdot \$250$ requires the percentage be converted to a decimal before multiplying.

To convert from percentage to decimal, **divide the decimal by 100** AND remove the "%" sign OR the quick way to divide the decimal by 100 is to **move the decimal point 2 places to the left** and remove the "%" sign.

To convert from decimal to percentage, **multiply the decimal by 100** AND write the "%" sign OR the quick way to multiply by 100 is to **move the decimal point 2 places to the right** and write the "%" sign.

Guided Examples	Practice
1) 3.9% converts to the decimal 0.039	1) Convert 200% to a decimal
2) 7% converts to the decimal 0.07	2) Convert 6.1% to a decimal
3) the decimal 0.45 converts to 45%	3) Convert 8% to a decimal
4) the decimal 0.071 converts to 7.1%	4) Convert 0.079 to a percentage
5) the decimal 23 converts to 2300%	5) Convert 4.1 to a percentage
6) What is 8.2% of 60?	6) What is 5.4% of 230?
First, convert the percent to a decimal 8.2% becomes 0.082. Next, multiply $0.082 \cdot 60 = 4.92$.	
	7) What is 38% of 750?
	8) You deposit $\$500$ in a bank or credit union account with an annual (simple) interest rate of 4%. How much interest do you earn after one year? For now, multiply the deposit and the interest rate (HINT: First convert the interest percent to a decimal).

Name:_____Date:_____

Instructor:_____Section:_____

Objective 4.R.4 – Solve linear equations using addition and multiplication

Solving one step equations is a required calculation for working with loan basics. Your calculations will involve the principal (amount of money you borrow), the interest you are charged and your efforts to gradually pay down the principal.

Solving these linear equations is the task of isolating (solving for) the variable x. To isolate the variable on one side of the equal sign, you must keep in mind the following:

- start on the side with the variable
- look to do the opposite operation – always start with addition or subtraction
- what you do to one side of the equal sign, you must do to the other side of the equal sign.
- then look to do the opposite operation – multiplication or division to finish

Guided Examples	Practice
Solve the linear equation. 1) $5x - 1 = 19$ $5x - 1 = 19$ start on the left side(side with x) $\underline{+1 \quad +1}$ the opposite of -1 is +1 and do this to both sides $\dfrac{5x}{5} = \dfrac{20}{5}$ the opposite of multiply by 5 is to divide by 5 and do this to both sides $x = 4$ 2) $3x + 5 = 35$ $3x + 5 = 35$ start on the left side(side with x) $\underline{-5 \quad -5}$ the opposite of +5 is -5 and do this to both sides $\dfrac{3x}{3} = \dfrac{30}{3}$ the opposite of multiply by 3 is to divide by 3 and do this to both sides. The result is $x = 10$	Use algebraic concepts to solve the following linear equation: 1) $4x + 2 = 26$ 2) $5x - 12 = 18$ 3) $3x - 15 = 12$

Objective 4.R.5 – Evaluate radical expressions

Using savings plan formulas, investment gains and rates or return require calculations using an understanding of roots and the radical symbol.

The $\sqrt{}$ symbol is called the radical symbol. A square (second) root is written as $\sqrt[2]{}$. In this case, the index number 2 is not written, so $\sqrt[2]{}$ is the same as $\sqrt{}$. There are other roots to consider. A cube (third) root is written as $\sqrt[3]{}$, a fourth root is written as $\sqrt[4]{}$ and a fifth root is written as $\sqrt[5]{}$.

When you use the radical symbol, you have to ask yourself, "What same number times itself by the index number is equal to the value under the radical sign?

Let's do a couple of examples to become familiar with this notation and how to use it.

EX1: $\sqrt[2]{36}$. What two same numbers multiply to 36 ? We state the square root of $36 = 6$, because $6 \cdot 6 = 36$

EX2: $\sqrt[4]{81}$. What four same numbers multiply to 81 ? We state the fourth root of $81 = 3$, because $3 \cdot 3 \cdot 3 \cdot 3 = 81$.

Now, most values are not perfect squares. When this happens, your answer will be a decimal approximation. For these approximations, a calculator can be a useful utility.

EX3: $\sqrt{18} \approx 4.24$ (approximate). This means that $4.24 \cdot 4.24$ is about 18 .

Guided Examples	Practice
Evaluate the radical expression:	Evaluate the radical expression:
1) $\sqrt{9}$ What two same numbers multiply to 9? The square root of $9 = 3$.	1) $\sqrt{144} =$
2) $\sqrt[3]{125} =$ What three same numbers multiply to 125? The cube root of $125 = 5$.	2) $\sqrt[3]{64} =$
3) $\sqrt[4]{256}$ What four same numbers multiply to 256? The fourth root of $256 = 4$.	Approximate the radical expression: 3) $\sqrt{24} =$
4) $\sqrt[5]{32}$ What five same numbers multiply to 32? The fifth root of $32 = 2$.	4) If you invest $\$3,000$ that grows to $\$8,400$ over four years, what is the percent increase? The radical expression looks like this: $\sqrt[4]{\dfrac{8400}{3000}} =$

Name:_____Date:_____

Instructor:_____Section:_____

Chapter 5 Statistical Reasoning

Learning Objectives

Objective 5.R.1 – Read and interpret graphs

Objective 5.R.1 – Read and Interpret Graphs

A **bar graph** is a two dimensional visual diagram in which the numerical counts of a categorical variable are represented by the height (or length) of rectangles of equal width. This visual display is used to compare characteristics of two types of data groups.

There are several items to consider when reading, analyzing and interpreting a bar graph. These are the graph title, the axes labels, scale and the height of the bars. This attention to graph details will improve the accuracy of your interpretation of the data presented in the graph.

First, locate the horizontal line on the graph (*x*-axis) that runs along the bottom of the bar graph. Typically, this is the location of information describing the data group for each bar.

Next, locate the vertical line on the left side of the bar graph (*y*-axis). This is the location of information describing the numerical value of the data group. The scale of this numerical data (hundreds, millions) is critical to your interpretation and analysis and is almost limitless in how it is presented.

Finally, your ability to read and interpret this correlating information is instrumental in your analysis. In your analysis of the data presented in the bar graph, you will correlate (or match) the data on the horizontal line (*x*-axis) with the data on the vertical line (*y*-axis).

Guided Example

Analyze the below bar graph:

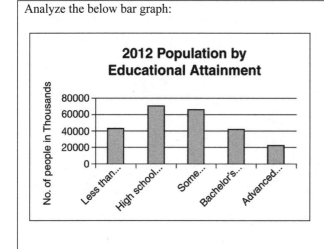

1) Which level of education has the highest population? High school graduate

2) Which two levels of education have approximately the same level of education? Less than HS and Bachelor's

3) What is the approximate number (in thousands) of people with advanced degrees? 21,000 (in thousands)

Practice

Analyze the below bar graph:

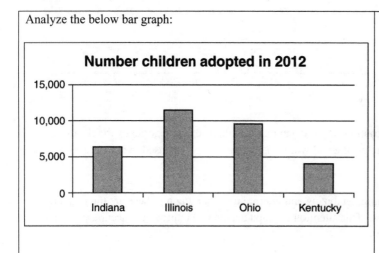

Number children adopted in 2012

1) Which state had the highest number of children adopted in 2012?

2) Which state had fewer than 5000 children adopted in 2012?

3) Approximately how many children were adopted in the state of Ohio in 2012?

Name:_____Date:_____

Instructor:_____Section:_____

Chapter 6 Putting Statistics to work

Learning Objectives
Objective 6.R.1 – Calculate average Objective 6.R.2 – Find the square root of a number Objective 6.R.3 – Find the percent by solving an equation Objective 6.R.4 – Evaluate and approximate radical expressions

Objective 6.R.1 – Calculate average

A measure of central tendency is the calculation of a value that attempts to numerically describe or identify the central position within a set of data. You may be familiar with the mean (also called the average) as a measure of central tendency. Additionally, we will explore the median and the mode.

The mean, median and mode are all measures of central tendency, but they are calculated differently and under varied circumstances, some measures of central tendency are more appropriate and accurate to use in characterizing, analyzing and interpreting your data.

Measure of central tendency	Description
Mean	Also called **average value**. The mean is calculated by adding up all the scores and dividing the total by the number of data items in the illustration.
Median	Is the **middle value**. The median represents the value in the exact middle of a number of data items when the data are arranged in ascending (lowest to highest) order or descending (highest to lowest) order. In this case, half of the data items are above the median and half are below the median. NOTE: If the data items total to an even number, then the median is the average of the two middle terms when they are written in order.
Mode	Is the **most common value.** The mode is identified as the value with the highest frequency of all data items in the illustration. To clarify, it is the most common value or the value that appears most often.

Guided Examples

Practice

| Calculate the measures of central tendency – mean, median and mode for the data provided:

1) A study of newborn birth weights at the local hospital on a specific day showed that 7 babies were born and their weights (in pounds) were: $7,5,6,9,12,6,11$.

Mean $= \dfrac{7+5+6+9+12+6+11}{7} = \dfrac{56}{7} = 8$ pounds | Calculate the measures of central tendency – mean, median and mode for the data provided:

1) A study of the age of children (in years) in a small daycare show the ages to be: $2,5,7,4,2,2,6$.

a) Mean $=$ |

| For median, the original sequence becomes $5,6,6,7,9,11,12$ when the values are arranged in an ascending numerical sequence. Since 7 is the middle term of this sequence, the median is 7 pounds.

NOTE: If the data items are $6,6,7,9,11,12$, the data items total an even number, so you must average the two middle terms. With these data items, the median is $\dfrac{7+9}{2} = \dfrac{16}{2} = 8$.

Mode $= 6$ pounds. The most common value is 6. | b) Median $=$

c) Mode $=$ |

Name:_____ Date:_____

Instructor:_____ Section:_____

Objective 6.R.2 – Find the square root of a number

Often times, the collection of data can be described as clustering closely together or far apart. When you get a cup of your favorite coffee from your local coffee shop, is it consistently the same delicious taste or does the taste vary greatly from visit to visit? Your response is known as a measure of variation.

The measure of variation describes how closely grouped or scattered a collection of data can be. The three measures of variation are the range, the variance and the standard deviation. To accurately work with these measures of variation requires us to correctly find the square root of a number.

The $\sqrt{}$ symbol is called the radical symbol. A square (second) root is written as $\sqrt[2]{}$. In this case, the index number 2 is not written, so $\sqrt[2]{}$ is written as $\sqrt{}$. When you use the radical symbol, you have to ask yourself, "What same number times itself, the index number of times, is equal to the value of the radicand (the value under the radical sign)?

Let's do an example to become familiar with this notation and how to use it.

EX1: $\sqrt[2]{25} = \sqrt{25}$. What two same numbers multiply to 25? We state the square root of $25 = 5$, because $5 \cdot 5 = 25$

NOTE: In this section, we are only working with square roots, so your precise answer or approximation should be a non-negative (a positive) value.

Guided Examples   Practice

Calculate the square root of the following examples:	Calculate the square root of the following examples:
1) $\sqrt{16} = 4$ Because $4 \cdot 4 = 16$.	1) $\sqrt{9} =$
2) $\sqrt{144} = 12$ Because $12 \cdot 12 = 144$.	2) $\sqrt{64} =$
3) $\sqrt{28} \approx 5.29$ The $\sqrt{28}$ will be an approximation because there are no two exact numbers that multiply to 28 (28 is not a perfect square). A calculator is helpful when approximating square roots. To check, $5.29 \cdot 5.29 \approx 28$.	3) Approximate the $\sqrt{48} =$

31

Objective 6.R.3 – Find the percent by solving an equation

In this section, you will be challenged to calculate and interpret numerical perspective across a variety of math modeling applications. In order to accurately calculate, analyze and translate your results, we should briefly discuss numerical perspectives using a general format.

The general form builds a proportion using two ratios set equal to each other, then solve by cross multiplication. A proportion is a label given to a statement that two ratios are equal. It can be written as:

$\frac{a}{b} = \frac{c}{d}$. When two ratios are equal, then the cross products of the ratios are equal. For this general form, that means that $a \cdot d = b \cdot c$. The numerator of the first fraction is multiplied by the denominator of the second fraction and set equal to the denominator of the first fraction multiplied by the numerator of the second fraction.

Specifically, for math modeling and to advance numerical perspective, a specific proportion is created as follows:

- $\frac{is}{of} = \frac{\%}{100}$.

We can use this format to answer the question, "What is 30% of 60 ?" The challenge will be to put the values in the correct location prior to your calculation and solve for the unknown (x). We always have to work with three values:

1) The 100 never changes location, the 30 is linked with the % symbol and the 60 is linked with the word *of*. The result is $\frac{x}{60} = \frac{30}{100}$. We now have to solve for x. Use cross products to get $100x = 60 \cdot 30$. So, $100x = 1,800$. Lastly, divide both sides by 100 to get $x = 18$.

Guided Examples	Practice
Solve the math modeling problems for x.	Solve the math modeling problems for x.
1) What is 45% of 120? ?	1) What is 20% of 300?
$\frac{x}{120} = \frac{45}{100}$ Next, cross multiplication results in $100x = 120 \cdot 45$. By multiplying the right side, you get $100x = 5400$. Last, divide both sides by 100, to get $x = 54$ What three values are we working with? The 100 never changes location, the 45 is linked with the % symbol and 120 is linked with the word *of*.	
	2) 60 is what percent of 200 ?
2) 30 is what percent of 150 ?	3) 40% of what number is 50 ?
$\frac{30}{150} = \frac{x}{100}$ Next, $30 \cdot 100 = 150 \cdot x$. $3,000 = 150 \cdot x$. $20 = x$. The answer is $x = 20\%$. What three values are we working with? The 100 never changes location, the 30 is linked with the word *is* and the 150 is linked with the word *of*.	

Name:_____Date:_____

Instructor:_____Section:_____

Objective 6.R.4 – Evaluate and approximate radical expressions

When you analyze a statistical study, you will be expected to determine the statistical significance of the numerical data collected. To precisely calculate and quantify statistical significance requires you to evaluate and approximate radical expressions.

Specifically, the radical symbol ($\sqrt{}$) is used when evaluating margin of error and the confidence interval. The mathematical relationship is as follows:

$$\textbf{margin of error} \approx \frac{1}{\sqrt{n}}$$

Guided Examples   Practice

Approximate the margin of error using the formula:	Approximate the margin of error using the formula:
1) $\dfrac{1}{\sqrt{250}} \approx 0.063$ To calculate an approximate answer will require the use of calculator. The formula means $1 \div \sqrt{250}$ and can be entered into a calculator. The result is $1 \div \sqrt{250} = 1 \div 15.81 \approx 0.063$. 2) $\dfrac{1}{\sqrt{500}} \approx .044$ To calculate an approximate answer will require the use of calculator. The formula means $1 \div \sqrt{500}$ and can be entered into a calculator. The result is $1 \div \sqrt{500} = 1 \div 22.36 \approx .045$. 3) Calculate the margin of error for a survey of 600 people $(n = 600)$. Use the formula **margin of error** $\approx \dfrac{1}{\sqrt{n}}$. When you substitute 600 for n in the formula, the formula means $1 \div \sqrt{600}$ and can be entered into a calculator. The result is $1 \div \sqrt{600} = 1 \div 24.49 \approx .041$.	1) $\dfrac{1}{\sqrt{300}} \approx$ 2) $\dfrac{1}{\sqrt{50}} \approx$ 3) Calculate the margin of error for a survey of 200 people $(n = 200)$. Use the formula **margin of error** $\approx \dfrac{1}{\sqrt{n}}$.

Chapter 7 Probability: Living with the Odds

Learning Objectives
Objective 7.R.1 – Evaluate an exponential expression Objective 7.R.2 – Multiply real numbers Objective 7.R.3 – Write or evaluate algebraic expressions using exponents

Objective 7.R.1 – Evaluate an exponential expression

To develop an accurate understanding of the fundamentals of probability, we will acquire the ability to evaluate an exponential expression. Specifically, we will advance our understanding, interpretation and calculation of the Fundamental Counting Principle and combining probabilities.

These concepts require us to further explore and precisely apply exponents. In the general form, a^n , a is the base number and n is the exponent. An exponent is a number that tells how many times the base number is multiplied (a factor).

For example, 5^3 indicates that the base number 5 is multiplied 3 times. To determine the value of 5^3 , multiply $5 \cdot 5 \cdot 5$ which equals 125 $(5 \cdot 5 \cdot 5 = 125)$.

When combining probabilities, fractions are frequently used. The use of fractions in an exponential expression is similar to the previous example.

For example, $\left(\dfrac{3}{4}\right)^2$ indicates that the numerator (3) AND the denominator (4) are multiplied 2 times. To continue,

the value of $\left(\dfrac{3}{4}\right)^2 = \dfrac{3 \cdot 3}{4 \cdot 4} = \dfrac{9}{16}$.

If you are multiplying fractions that do not use an exponent, your calculation process is similar – multiply across the numerator and multiply across the denominator. A common denominator is not required when you multiply

fractions. If necessary, simplify your answer. For example, $\left(\dfrac{3}{5}\right) \cdot \left(\dfrac{2}{4}\right) \cdot \left(\dfrac{1}{3}\right) = \dfrac{3 \cdot 2 \cdot 1}{5 \cdot 4 \cdot 3} = \dfrac{6}{60}$. This further reduces by

finding and dividing by the GCF (Greatest Common Factor), so the final answer looks like $\dfrac{6}{60} = \dfrac{6 \div 6}{60 \div 6} = \dfrac{1}{10}$.

Guided Examples	Practice
Calculate the following exponential expressions: 1) $4 \cdot 2 \cdot 3 = 24$ Consider the order of operations with this problem. Since the operations are all multiplication, just multiply the numbers from left to right to get 24 .	Calculate the following exponential expressions: 1) $4 \cdot 2 \cdot 3 \cdot 3 =$

2) You are eating at a cafeteria that offers 3 types of salads, 2 types of meats, 5 different vegetables and 4 different desserts. How many ways can you choose a different meal?

$3 \cdot 2 \cdot 5 \cdot 4 = 120$ different ways

3) $\left(\dfrac{4}{5}\right)^3 =$

To solve, remember the exponent (3), is distributed to both the numerator (4) and denominator (5). The result is

$$\left(\dfrac{4}{5}\right)^3 = \dfrac{4 \cdot 4 \cdot 4}{5 \cdot 5 \cdot 5} = \dfrac{64}{125}$$

4) $\left(\dfrac{6}{5}\right) \cdot \left(\dfrac{5}{4}\right) \cdot \left(\dfrac{4}{3}\right) =$

Multiply straight across the numerator and straight across the denominator (no common denominator is required when you multiply fractions).

$$\left(\dfrac{6}{5}\right) \cdot \left(\dfrac{5}{4}\right) \cdot \left(\dfrac{4}{3}\right) = \dfrac{6 \cdot 5 \cdot 4}{5 \cdot 4 \cdot 3} = \dfrac{120}{60} = 2$$

2) You look in your closet and have 4 tops, 5 pants and 12 pair of shoes. How many different outfits can you make (assuming all items match)?

3) $\left(\dfrac{5}{6}\right)^2 =$

4) $\left(\dfrac{5}{10}\right) \cdot \left(\dfrac{4}{9}\right) \cdot \left(\dfrac{3}{8}\right) =$

Objective 7.R.2 – Multiply real numbers

When working with the Law of Large Numbers, as it relates to probability, you will explore the concept of expected value. Expected value is a value you would "expect" to achieve if you could repeat a process an "infinite" number of times (Law of Large Numbers).

This concept leads to decision theory that is most often used for business decision making. The expected value model is used in a situation where monetary gains and losses will occur. After calculating the expected value, the optimal choice would be the alternative that makes the most money (has the highest expected value).

The expected value formula requires the multiplication of real numbers. In words, to compute the expected value you would multiply the monetary payoff for each alternative outcome by the probability of the event occurring. The formula is:

$$E(v) = (\text{outcome\#1} \cdot \text{probability}) + (\text{outcome\#2} \cdot \text{probability}) + (\text{outcome\#3} \cdot \text{probability})...$$

Guided Examples	Practice
Multiply the real numbers:	Multiply the real numbers:
1) $5 \cdot \dfrac{3}{4} =$ There are two ways to multiply these numbers. Choose the method that works best for you: a) Change the first number to a fraction (denominator $=1$), then multiply the numerators and the denominators. This becomes $\dfrac{5}{1} \cdot \dfrac{3}{4} = \dfrac{5 \cdot 3}{1 \cdot 4} = \dfrac{15}{4} = 3.75$ b) Use order of operations (PEMDAS) and calculate the problem left to right. The whole number (5) is multiplied to the numerator (3) of the fraction, then the results is divided by the denominator (4) of the fraction. This becomes $5 \cdot 3 \div 4 = 3.75$ 2) $\left(12 \cdot \dfrac{2}{3}\right) + \left(15 \cdot \dfrac{1}{3}\right) =$ a) Convert the first number to a fraction (denominator $=1$), then multiply across the numerator and across the denominator. $\left(\dfrac{12}{1} \cdot \dfrac{2}{3}\right) + \left(\dfrac{15}{1} \cdot \dfrac{1}{3}\right) = \left(\dfrac{12 \cdot 2}{1 \cdot 3}\right) + \left(\dfrac{15 \cdot 1}{1 \cdot 3}\right) = \left(\dfrac{24}{3}\right) + \left(\dfrac{15}{3}\right) = 8 + 5 = 13$ b) Using the order of operations (PEMDAS), work in the first parentheses first, then the second set of parentheses. Combine the resulting two numbers. $(12 \cdot 2 \div 3) + (15 \cdot 1 \div 3) = (24 \div 3) + (15 \div 3) = 8 + 5 = 13$	1) $8 \cdot \dfrac{4}{5} =$ 2) $\left(14 \cdot \dfrac{2}{5}\right) + \left(10 \cdot \dfrac{3}{5}\right) =$

3) You expect your grandmother will send you money for your birthday. Based on past experience, you figure there is a $\frac{3}{10}$ chance your grandmother will send you $20 and a $\frac{7}{10}$ chance she will send you $50. What is your expected value for this event? The formula looks like:

$$E(v) = \left(20 \cdot \frac{3}{10}\right) + \left(50 \cdot \frac{7}{10}\right)$$

a) $E(v) = \left(\frac{20}{1} \cdot \frac{3}{10}\right) + \left(\frac{50}{1} \cdot \frac{7}{10}\right) = \left(\frac{60}{10}\right) + \left(\frac{350}{10}\right) = 6 + 35 = \$41$

b) $E(v) = (20 \cdot 3 \div 10) + (50 \cdot 7 \div 10) = 6 + 35 = \$41$

This means that if your grandmother did this for many, many years – on average you would receive $41 on your birthday.

3) Your community center is giving away raffle cards (no charge to you) and every card is a winner. There is a $\frac{9}{10}$ chance you will win $2 and a $\frac{1}{10}$ chance you will win $10. If you were to receive a large, large quantity of these tickets, what is your expected value for this event?

Objective 7.R.3 – Write or evaluate algebraic expressions using exponents

When exploring and studying probability concepts, there are some basic counting techniques employed to discover possible outcomes for a specific situation. In this section, we try to give precise mathematical meaning to questions such as, "How many ways…?, "What are the chances…? and "What is the likelihood…?".

Specifically, these questions make take the form:

- How many ways can you create a 4 digit ATM pin number?
- What are the chances of selecting an all women committee if the group consists of men and women?
- What is the likelihood of rolling a pair of $6's$ in a dice game?

To answer these counting questions, we will explore the concepts of evaluating mathematical expressions using exponents and factorial notation.

First, evaluating mathematical expressions using exponents is helpful when studying products where the same factor may occur more than once. For example, write the following expression using exponents:

- $15 \cdot 15 \cdot 15 \cdot 4 \cdot 4 \cdot 4 \cdot 4 \cdot 4 = (15)^3 \cdot (4)^5$

Please note that when working with exponents, the operation is multiplication. Then, count the number of times a factor repeats – the factor 15 repeats 3 times (3 is the exponent) and the factor 4 repeats 5 times (5 is the exponent).

Next, the use of factorials is a help when exploring and mastering the concepts of permutations and combinations. There is a specialized and specific symbol and pattern associated with factorials.

- Factorial symbol is !. The ! is a shorthand way to denote the multiplication of consecutive positive integers in descending order.
- The pattern is demonstrated as $4! = 4 \cdot 3 \cdot 2 \cdot 1 = 24$
  Again, the pattern is demonstrated as $6! = 6 \cdot 5 \cdot 4 \cdot 3 \cdot 2 \cdot 1 = 720$

NOTE: As you have noticed, the calculation is straightforward and the numbers get large quickly. There is a factorial button (!) on your calculator to help you with large calculations.

Guided Examples   Practice

Write the following expressions using exponents:	Write the following expressions using exponents:
1) $5 \cdot 5 \cdot 5 \cdot 8 \cdot 8 \cdot 8 \cdot 8 =$ Is the operation multiplication? If so, you can use exponents to rewrite this expression. Then, count the number of times a factor repeats – the factor 5 repeats 3 times (3 becomes the exponent) and the factor 8 repeats 4 times (4 becomes the exponent). $5 \cdot 5 \cdot 5 \cdot 8 \cdot 8 \cdot 8 \cdot 8 = (5)^3 \cdot (8)^4$	1) $2 \cdot 2 \cdot 2 \cdot 2 \cdot 7 \cdot 7 \cdot 7 =$

2) $(3)(3)(12)(12)(12)(12)(12) =$

The parentheses indicate multiplication and do not affect your decision.

$(3)(3)(12)(12)(12)(12)(12) = (3)^2 \cdot (12)^5$

Calculate the value of the expressions involving factorials.

3) $5! =$
$5! = 5 \cdot 4 \cdot 3 \cdot 2 \cdot 1 = 120$

4) $(12-9)! =$
Recall order of operations (PEMDAS) and work in the parentheses first. So, $(12-9)! = (3)! = 3 \cdot 2 \cdot 1 = 6$.

5) $5! - 4! =$
Calculate each factorial value first, then subtract. So, $5! - 4! = 120 - 24 = 96$.

6) $\dfrac{6!}{3!} =$
Calculate each factorial value first, then divide. So, $\dfrac{6!}{3!} = \dfrac{720}{6} = 120$.

7) You own a 4 digit bike lock that can use the single digits $0,1,2,3,4,5,6,7,8,9$ (10 numbers). How many ways can you create a different bike lock combination (if you can repeat numbers)?

You have 4 spaces and you can place any one of the 10 numbers in each space, so
$10 \cdot 10 \cdot 10 \cdot 10 = (10)^4 = 10,000 \text{ ways}$.

2) $(9)(9)(9)(4)(4) =$

Calculate the value of the expressions involving factorials.

3) $4! =$

4) $(10-5)! =$

5) $5! - 3! =$

6) $\dfrac{6!}{4!} =$

7) You open a new bank account and you want to set up your new 4 digit ATM pin number. Bank regulations state that you have to use the single digit numbers $0,1,2,3,4,5,6,7,8,9$. How many ways can you create a different ATM pin number (if you can repeat numbers)?

Name:_____ Date:_____

Instructor:_____ Section:_____

Chapter 8 Exponential Astonishment

Learning Objectives

Objective 8.R.1 – Graph exponential functions
Objective 8.R.2 – Solve exponential equations
Objective 8.R.3 – Evaluate logarithms

Objective 8.R.1 – Graph exponential functions

There are two basic growth patterns – linear growth and exponential growth. Linear activity is very common, but there are events that cannot be explained with linear math modeling. For example:

- In biology, microorganisms grow at an exponential rate until the nutrients are consumed.
- In physics, nuclear power activity is best explained with exponential rates.
- In finance, compound interest behaves in an exponential pattern.
- In computer technology, the processing power of computers (Moore's Law) can be described using an exponential pattern.
- In medicine, the way medicine and caffeine in your system wears off is best described by exponential decay.

Using your personal finances, if you receive a $500 wage increase each year, then this is an example of linear growth. If you receive a 10% wage increase each year then this is an example of exponential growth. Decay or decrease works in much the same way for both linear and exponential math modeling.

Our challenge this section will be to examine and interpret linear growth/decay and exponential growth/decay as it relates to graphs and real world applications.

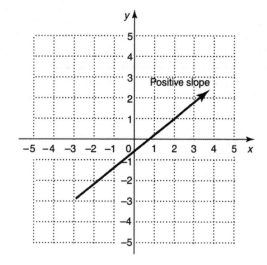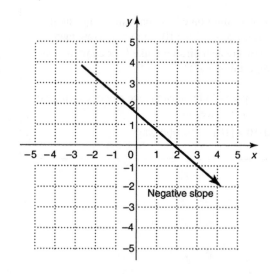

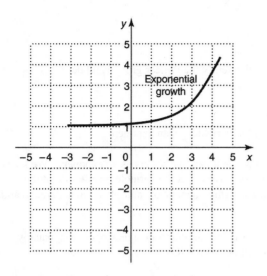

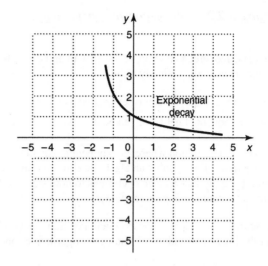

Guided Examples   Practice

Determine if the written example or graph is an example of linear growth, linear decay, exponential growth or exponential decay:	Determine if the written example or graph is an example of linear growth, linear decay, exponential growth or exponential decay:
1) The student enrollment at your College or University has increased by 350 people each year for the past 6 years. This is a yearly numerical increase, so this is linear growth.	1) You have noticed that around the holidays, the price of gasoline increases $.10 per day for many days.
2) You have started a healthy living program. You are starting to see the benefits and have lost 2 pounds a week for the past 5 weeks. This is a weekly numerical decrease, so this is linear decay.	2) The value of your car decreases by 10% each year for many years.
3) You are a fiscally responsible person and have invested some of your money in a bank account earning a 2% rate of interest, compounded monthly. This is a monthly percent gain, so this is exponential growth.	3) As the calendar moves from summer to fall to winter, the days get shorter by about 3 minutes each day.
4) Because of the difficult economic times, homes in certain parts of the United States decreased in value by 7% a year for 6 years. This is a yearly percent decline, so exponential decay.	4) The federal government keeps a statistic called the CPI (Consumer Price Index). This index measures inflation from year to year. On average, inflation has increased by about 2% each year for the past 10 years.

Objective 8.R.2 – Solve exponential equations

You have discovered that exponential growth and decay behaves differently than linear growth and decay. This difference occurs algebraically as well:

- Linear behavior can be expressed as $y = 2x$.
- Exponential behavior can be expressed as $y = 2^x$, where the variable x is an exponent.

This leads to a definition where an exponential equation is one in which a variable occurs as an exponent. We know from experience that to solve an algebraic equation for x requires you have to get the variable on one side of the equation by applying opposite operations to both sides of the equation (the opposite of multiplication is division and the opposite of addition is subtraction).

Because the variable x is in the exponent position, we will have to become familiar with a new operation, namely, taking the log of both sides. The study of logarithmic functions will occur in 8.R.3.

The opposite (technically the "inverse") of exponentials are logarithms, so to isolate the variable x in the exponent position requires us to take the log of both sides of the equation. To solve most exponential equations requires a four step process:

1. Isolate the exponential variable.
2. Take the logarithm (log) of both sides
3. Solve for the variable
4. Use a calculator to solve

For now, the most useful log rule states that exponents inside a log become multipliers in front of the log. Let's look at example:

Example	Explanation
Solve $5^x = 7$ to the nearest thousandth	Because x is an exponent, we need to use logs.
$\log 5^x = \log 7$	Take the log of both sides
$x \log 5 = \log 7$	The log rule states that the exponent becomes a multiplier – move the exponent to the front of the log.
$\dfrac{x \log 5}{\log 5} = \dfrac{\log 7}{\log 5}$	Now, we can isolate the variable x, by dividing both sides by $\log 5$.
$x = \dfrac{\log 7}{\log 5}$	The $\log 5$ on the left side cancels (reduces to 1), so the variable x is isolated.
$x = \dfrac{0.845}{0.699}$	You will need to locate the log button on your calculator. So, $\log 7 \approx 0.845$ and $\log 5 \approx 0.699$.
$x = \dfrac{0.845}{0.699} \approx 1.209$	To evaluate, $5^{1.209} \approx 7$. We have solved for x when it is in the exponent position.

Guided Examples	Practice
Solve the following for x to the nearest thousandth: 1) $7^x = 60$ This is an exponential equation, so we have to work with logs. First, take the log of both sides so $\log 7^x = \log 60$. Next, the log rule states that the x can move to the front of the log as a multiplier, so $x \log 7 = \log 60$. Next, isolate x by dividing both sides by $\log 7$, so you get $x = \dfrac{\log 60}{\log 7}$. We need a calculator to solve for x. The result is $x = \dfrac{1.778}{0.845} \approx 2.104$. To check, $7^{2.104} \approx 60$. 2) $9^x = 100$ This is an exponential equation, so we have to work with logs. First, take the log of both sides so $\log 9^x = \log 100$. Next, the log rule states that the x can move to the front of the log as a multiplier, so $x \log 9 = \log 100$. Next, isolate x by dividing both sides by $\log 9$, so you get $x = \dfrac{\log 100}{\log 9}$. We need a calculator to solve for x. The result is $x = \dfrac{2.000}{0.954} \approx 2.096$. To check, $9^{2.096} \approx 100$.	Solve the following for x to the nearest thousandth: 1) $3^x = 40$ 2) $12^x = 75$

Objective 8.R.3 – Evaluate logarithms

In our next undertaking, we will investigate the use of logarithmic scales. Logarithmic scales exist when activities revolve around orders of magnitude. These areas include the magnitude scale for earthquakes, the loudness of sound measured in decibels and the pH scale for acidity.

This will cause us to examine the method to convert logarithms to exponential form. This entails transforming the logarithmic form $y = \log_b x$ if and only if $b^y = x$. Let's look at a challenge:

Evaluate $\log_3 9$. Rewrite the problem as $\log_3 9 = y$. The pattern to transform from log form to exponential form results in $3^y = 9$. There are two methods to find the value for y:

- Use the log form we just discovered in 8.R.2. The result is $\log 3^y = \log 9$. Then the y moves to the front of the log as a multiplier $y \log 3 = \log 9$. Divide both sides by $\log 3$ to isolate the variable y. So $y = \dfrac{\log 9}{\log 3}$.

  Finally, using a calculator, $y = \dfrac{0.954}{0.477} \approx 2$.

- You ask yourself the question, "3 to what value exponent (y) equals 9 ?" $y = 2$. To check, $3^2 = 9$.

Guided Examples

Practice

Evaluate the following logarithms:	Evaluate the following logarithms:
1) $\log_5 25$	1) $\log_2 32$
Rewrite the problem to $\log_5 25 = y$. Then, use the pattern to transform the log form to exponential form, so $5^y = 25$. Finally, ask yourself the question "5 to what value exponent equals 25 ?" $y = 2$. To check $5^2 = 25$.	
2) $\log_4 64$	2) $\log_{10} 1000$
Rewrite the problem to $\log_4 64 = y$. Then, use the log rule to transform the log form to exponential form, so $4^y = 64$. Finally, ask yourself the question "4 to what value exponent equals 64 ?" $y = 3$. To check $4^3 = 64$.	
3) $\log_3 81$	3) $\log_5 125$
Rewrite the problem to $\log_3 81 = y$. Then, use the pattern to transform the log form to exponential form, so $3^y = 81$. Finally, ask yourself the question "3 to what value exponent equals 81 ?" $y = 4$. To check $3^4 = 81$.	

Chapter 9 Modeling Our World

Learning Objectives

Objective 9.R.1 – Identify and plot points on a coordinated plane
Objective 9.R.2 – Graph equations in the rectangular coordinate system
Objective 9.R.3 – Decide whether an ordered pair is a solution of a system of linear equations in two   variables
Objective 9.R.4 – Find the slope of a line
Objective 9.R.5 – Write equivalent exponential and logarithmic equations

Objective 9.R.1 – Identify and plot points on a coordinated plane

We will begin to explore the graphic and numeric relationship between values. If these values are related by a function, they are considered variables because the values can change. For now, our focus will be on the graphic representation of how these variables relate.

A function describes how a dependent variable changes with respect to one or more independent variables. The related variables are often written in the form of an ordered pair (independent variable, dependent variable).

Functions are demonstrated graphically using a coordinate plane. The organization of the coordinate plane is as follows:

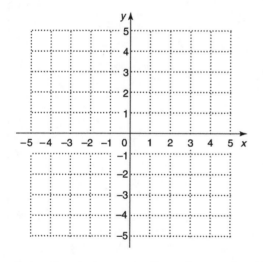

Points (location) on a coordinate plane are described by using an ordered pair as well. The ordered pair is in the form (x, y). To graph a point (identify a location), you start at the origin identified by the ordered pair $(0, 0)$. Due to the ordered pair format (x, y), the x coordinate value has you move right (positive) or left (negative) before the y coordinate has you move up (positive) or down (negative).

Guided Examples	Practice

Locate and graph the following points:

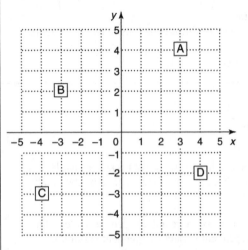

Identify the ordered pair values (x, y) for the given points:

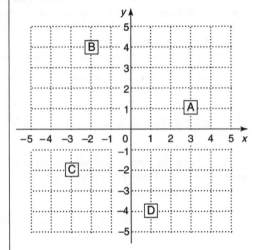

Point A has the ordered pair coordinates $(3,4)$ because from the origin, you travel $+3$ units (3 units right) along the *x axis direction*, then $+4$ units (4 units up) in the *y axis direction*.

Point B has the ordered pair $(-3,2)$ because from the origin, you travel -3 units (3 units left) along the *x axis direction*, then $+2$ units (units up) in the *y axis direction*.

Point C has the ordered pair $(-4,-3)$ because from the origin, you travel -4 units (4 units left) along the *x axis direction*, then -3 units (3 units down) in the *y axis direction*.

Point D has the ordered pair $(4,-2)$ because from the origin, you travel $+4$ units (4 units right) along the *x axis direction*, then -2 units (2 units down) in the *y axis direction*.

Point A: _____

Point B: _____

Point C: _____

Point D: _____

Objective 9.R.2 – Graph equations in the rectangular coordinate system

In the previous section, we plotted points in the rectangular coordinate system. A prevalent and advantageous way of representing functions is to use an equation to combine many points. Equations can be an excellent mathematical instrument when creating, examining and interpreting mathematical models.

A linear function is represented by a straight line, and as such, indicates a constant rate of change. The rate of change does not change throughout the graph of the line and indicates a rate that describes how one quantity changes in relation to another quantity.

To graph linear equations in the rectangular coordinate system, we refer to the linear formula $y = mx + b$. In words, the combined result is:

$$y = mx + b$$

$$\text{total value} = \text{rate of change} \cdot \text{variable} + \text{starting point}$$

If x is the independent variable and y is the dependent variable, then the rate of change can be expressed using the formula:

$$\text{rate of change} = \frac{\text{change in y}}{\text{change in x}}$$

NOTE: For non-vertical lines, the starting point is always on the y axis because it is the value of y when $x = 0$. Vertical lines create an exception. In this case, the starting point may not be on the y axis.

Guided Examples	Practice

Using the following story, put the data into the $y = mx + b$ format, then graph.	Using the following story, put the data into the $y = mx + b$ format, then graph.
1) There is no snow on the ground (starting point $= 0$). A snowstorm is on the way and forecast to snow at a constant rate of 2 inches every 1 hour. So, $y = mx + b$ becomes $y = \dfrac{2}{1}x + 0$ 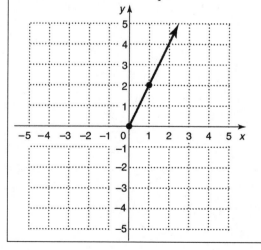	1) Your area is in the middle of a drought. Your local lake is now 5 inches below normal. Several rain storms are on the way and they are expected to increase the lake level by 2 inches every 3 days. 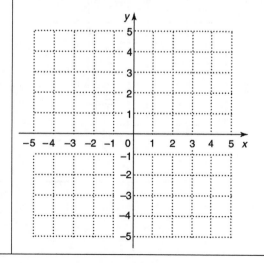

2) Because of recent rain, the local river is already 2 feet above flood stage and a new rain storm is on the way. The water level of the river is expected to rise 1 foot every 3 hours.

So, $y = mx + b$ becomes $y = \dfrac{1}{3}x + 2$

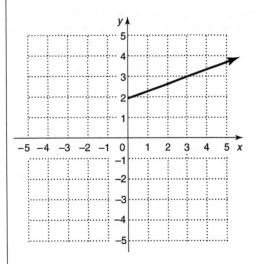

2) The world champion hot dog eater can eat 1 hot dog every 2 seconds. He is starting the contest (start point = 0).

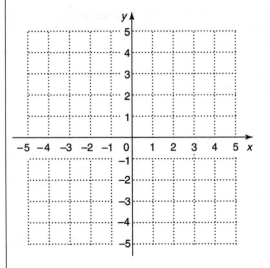

3) You start the holiday season 5 pounds overweight. Your goal is healthy living and a fitness program so you lose 2 pounds every 1 week.

So, $y = mx + b$ becomes $y = \dfrac{-2}{1}x + 5$.

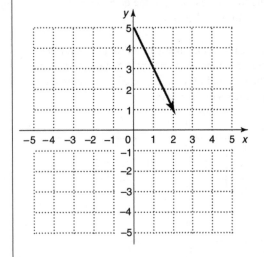

Objective 9.R.3 – Decide whether an ordered pair is a solution of a system of linear equations in two variables

Our algebraic challenge in this section deals with a solution that is provided to us in the form of an ordered pair (x, y). We need to decide if the information provided is accurate. Are you a skeptical person? Do you believe everything you are told or do you challenge items you read, see or hear? This is the decision making challenge that you are presented with in this section.

This involves taking the x value of the provided solution and substituting it for x in the specified equation, then taking the y value of the provided solution and substituting it for y in the specified equation. Complete the required algebraic operations then make your determination.

If the left side of the equation equals the right side of the equation, then the equation is balanced and it is determined that the values of the provided ordered pair (x, y) are a solution to the equation specified. If the left side does not equal the right side, then the equation is not balanced and it is determined that the values of the provided ordered pair (x, y) are not a solution to the equation specified. Let's look at two examples:

EX1: Is the ordered pair $(3, 7)$ a solution to the equation $2x + 5y = 41$?

The ordered pair $(3, 7)$ is in the format (x, y), so the x value is 3 and the y value is 7. You substitute these values into the specified equation $2x + 5y = 41$. After the substitution, the equation becomes $2 \cdot 3 + 5 \cdot 7 = 41$. Perform the algebraic operations to support your decision. The result is $6 + 35 = 41$, and finally $41 = 41$.

Does the left side balance with the right side? If yes, you can conclude that the ordered pair $(3, 7)$ is a solution to the equation $2x + 5y = 41$.

EX2: You are planning the itinerary for a Spring Break road trip with your friends. You are given TripTix by a travel agency. The agency calculates you can average 60 miles per hour and your destination is 1,210 miles away. The TripTix plan states that you can make the trip 18 hours. Do you believe their TripTix plan? Is their math correct?

You use your advanced algebra skills to determine the ordered pair $(18, 1210)$ and the equation $y = 60x + 0$ (the 0 is because you are starting the trip from your home).

The ordered pair $(18, 1210)$ is in the format (x, y), so the x value is 18 and the y value is 1210. You substitute these values into the specified equation $y = 60x + 0$. After the substitution, the equation becomes $1210 = 60 \cdot 18 + 0$. Perform the algebraic operations to support your decision. The result is $1210 \neq 1,080$.

Since the left side does not equal the right side, you can conclude that the TripTix information is not accurate.

Guided Examples	Practice
Apply algebraic operations to the given ordered pair and decide if the data given is a solution to the specified linear equation:	Apply algebraic operations to the given ordered pair and decide if the data given is a solution to the specified linear equation:

Apply algebraic operations to the given ordered pair and decide if the data given is a solution to the specified linear equation:

Guided Examples

1) Is the ordered pair $(1,8)$ a solution to the linear equation $9x - 2y = 7$?

The ordered pair $(1,8)$ is in the format (x, y), so the x value is 1 and the y value is 8. You substitute these values into the specified equation $9x - 2y = 7$. After substituting the values, the equation becomes $9 \cdot 1 - 2 \cdot 8 = 7$. Perform the algebraic operations to support your decision. The result is $9 - 16 = 7$, and finally $-7 \neq 7$.

The left side does not balance with the right side. No, the ordered pair $(1,8)$ is not a solution to the equation $9x - 2y = 7$.

2) There are 3 inches of snow on the ground and a big snowstorm is on the way. The storm is expected to dump 2 inches of snow per hour for 5 hours in the area. An emergency is declared if there are 13 inches or more of snow on the ground. If the snow storm is as bad as they predict, should an emergency be declared?

To describe this situation, use the ordered pair $(5,13)$ and the equation $y = 2x + 3$.

The ordered pair $(5,13)$ is in the format (x, y), so the x value is 5 and the y value is 13. You substitute these values into the specified equation $y = 2x + 3$. After the substitution, the equation becomes $13 = 2 \cdot 5 + 3$. Perform the algebraic operations to support your decision. The result is $13 = 13$.

Since the left side equals the right side, you can conclude that an emergency should be declared for the area.

Practice

Apply algebraic operations to the given ordered pair and decide if the data given is a solution to the specified linear equation:

1) Is the ordered pair $(3,10)$ a solution to the linear equation $10x - 2y = 10$?

2) You find a good deal on a car. The down payment is $2,000 with monthly payments of $230 for two years. The car dealer tells you that you need to budget $7,000 in order to pay for this deal. Is the dealer providing you with accurate information?

Use the ordered pair $(24, 7000)$ and the equation $y = 230x + 2,000$ to aid in your decision.

Name:_____Date:_____

Instructor:_____Section:_____

Objective 9.R.4 – Find the slope of a line

To conclude our exploration of linear equations on the rectangular coordinate plane, we focus on the slope of the line created by the linear equation. The slope of a line is the rate of change of the line between two points.

Geography is all about slope – do you travel uphill or downhill to get to campus? Slope is everywhere in architecture design – I have noticed that the slope of the roof of a home in New England is steep while the slope of the roof of a home in the South tends to be flat. Why is that?

There are many ways to think about the slope between two points. To combine these forms, slope is the rise of the run, the change in y divided by the change in x or the incline/decline of a line drawn between the two points.

The coordinates of the two points are specified using the ordered pair format. The first point is written using the general form (x_1, y_1) and the second point is written as (x_2, y_2). The general formula for slope is

$$slope = \frac{rise}{run} = \frac{\text{change in y (up or down)}}{\text{change in x (left or right)}} = \frac{y_2 - y_1}{x_2 - x_1}$$

EX1: What is the slope of the line segment connecting the points $(2,14)$ and $(5,20)$? First, you should place the coordinates for the points in their general form. So the first point $(2,14)$ is (x_1, y_1) and the second point $(5,20)$ is (x_2, y_2). Substitute into the slope equation to get $slope = \frac{y_2 - y_1}{x_2 - x_1} = \frac{20-14}{5-2} = \frac{6}{3} = 2$.

When you are asked to calculate slope, it makes no difference which point you call the first one and which point you call the second one. Since you are working with the same points, the slope of the line is the same regardless of which is the first point and which is the second point. To continue our example, let's determine the first point $(5,20)$ is (x_1, y_1) and the second point $(2,14)$ is (x_2, y_2). Substitute into the slope equation to get

$slope = \frac{y_2 - y_1}{x_2 - x_1} = \frac{14-20}{2-5} = \frac{-6}{-3} = 2$.

Guided Examples	Practice
Calculate the slope of the line segment connecting the coordinates of the two given points:	Calculate the slope of the line segment connecting the coordinates of the two given points:
1) $(3,8)$ and $(5,9)$	1) $(6,15)$ and $(8,20)$
First, you should place the coordinates for the points in their general form. The first point $(3,8)$ is (x_1, y_1) and the second point $(5,9)$ is (x_2, y_2). Substitute into the slope equation to get $slope = \dfrac{y_2 - y_1}{x_2 - x_1} = \dfrac{9-8}{5-3} = \dfrac{1}{2}$.	
2) $(8,14)$ and $(11,12)$	2) $(5,9)$ and $(8,7)$
First, you should place the coordinates for the points in their general form. So the first point $(8,14)$ is (x_1, y_1) and the second point $(11,12)$ is (x_2, y_2). Substitute into the slope equation to get $slope = \dfrac{y_2 - y_1}{x_2 - x_1} = \dfrac{12-14}{11-8} = \dfrac{-2}{3}$.	
3) $(-5,-9)$ and $(8,-15)$	
First, you should place the coordinates for the points in their general form. So the first point $(-5,-9)$ is (x_1, y_1) and the second point $(8,-15)$ is (x_2, y_2). Substitute into the slope equation to get $slope = \dfrac{y_2 - y_1}{x_2 - x_1} = \dfrac{-15-(-9)}{8-(-5)} = \dfrac{-15+9}{8+5} = \dfrac{-6}{13}$.	

Objective 9.R.5 – Write equivalent exponential and logarithmic equations

The goal of writing equivalent exponential and logarithmic equations is to solve for x when x is an exponent. This causes us to examine the method to convert logarithms to exponential form. The basic properties of logarithms lead us to explore two algebraic procedures. For now, our focus is on common logarithms (base 10). This can be written as $\log_{10} a^x = \log a^x$ and is identified on a calculator as the "log" button.

Our first general rule states that $\log_{10} a^x = x \log_{10} a$. The important portion of this rule is that the variable x can be moved from the exponent and placed in front of the term as a multiplier. For example,

Solve for x in the equation $2^x = 30$. Because the variable x is in the exponent place, we know to use logs to solve this problem. Take the log of both sides, so the result is $\log 2^x = \log 30$. Because of the rule, move the x to the front of the term, so $x \log 2 = \log 30$. To solve for x, divide both sides by $\log 2$. You get $x = \dfrac{\log 30}{\log 2} \approx \dfrac{1.47712}{0.30103} \approx 4.9069$.

The next general rule states requires us to recall from previous notes that $\log_b x = y$ if and only if $b^y = x$. Going forward, it applies that $y = 10^x$ and $y = \log_{10} x$ are inverses of each other.

Solve for x in the equation $4 \log_{10} = 20$. Because the variable x is in the logarithmic expression, the first step is to isolate $\log_{10} x$. This is done by dividing both sides by the factor 4. The result is $\log_{10} x = 5$. Next, we apply the general rule and get $10^5 = x$.

Guided Examples	Practice
Solve for x in the equation: 1) $4^x = 18$ Take the log of both sides, so the result is $\log 4^x = \log 18$ Because of the rule, move the x to the front of the term, so $x \log 4 = \log 18$. To solve for x, divide both sides by $\log 4$. You get $x = \dfrac{\log 18}{\log 4} \approx \dfrac{1.25527}{0.60206} \approx 2.0850$.	Solve for x in the equation: 1) $7^x = 53$

2) $3^x = 35$

Take the log of both sides, so the result is $\log 3^x = \log 35$
Because of the rule, move the x to the front of the term,
so $x \log 3 = \log 35$. To solve for x, divide both sides by
$\log 3$. You get $x = \dfrac{\log 35}{\log 3} \approx \dfrac{1.54407}{0.47712} \approx 3.2362$.

Solve for x in the equation:

3) $5 \log_{10} x = 15$.

The first step is to isolate the $\log_{10} x$. This is done by
dividing both sides by the factor 5. The result is
$\log_{10} x = 3$. After applying the general rule, $10^3 = x$ or
$x = 1000$.

4) $2 \log_{10} x = 5$.

The first step is to isolate the $\log_{10} x$. This is done by
dividing both sides by the factor 2. The result is
$\log_{10} x = \dfrac{5}{2}$. After applying the general rule, the result
is $10^{\frac{5}{2}} = x$ or $x = 10^{2.5}$.

2) $6^x = 220$

Solve for x in the equation:

3) $2 \log_{10} x = 4$

4) $5 \log_{10} x = 18$

Chapter 10 Modeling with Geometry

Learning Objectives

Objective 10.R.1 – Find the perimeter of a polygon or the circumference of a circle
Objective 10.R.2 – Find the area of a polygon or circle
Objective 10.R.3 – Find the volume of the geometric solid
Objective 10.R.4 – Find the missing sides of similar triangles
Objective 10.R.5 – Find the unknown side of a right triangle using the Pythagorean Theorem

---

Objective 10.R.1 – Find the perimeter of a polygon or the circumference of a circle

DISCLAIMER: Please note that shapes and measurements in this chapter may not be to scale.

The perimeter of any straight sided polygon is the total distance around the outside of the shape. The perimeter can be calculated by adding together the length of each side. This is true regardless of the type of polygon - a regular or an irregular shape. Perimeter is measured in one dimension, so the answer is in *units* like inches, feet, centimeters or meters.

For example, calculate the perimeter of the following irregular shape.

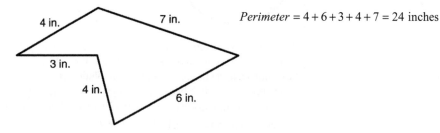

Perimeter $= 4 + 6 + 3 + 4 + 7 = 24$ inches

The perimeter of a circle is called the circumference. The formula for circumference is $C = \pi d$, where C is the circumference, π (*pi*)has an approximate value of 3.14 and d is the diameter (the distance across the circle with the path of the line going through the midpoint of the circle). To summarize, you can calculate the circumference of a circle by multiplying the diameter by π.

For example, calculate the circumference (distance around the circle) of the following circle.

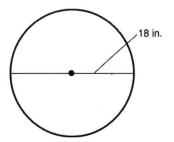

18 in.

Using the formula $C = \pi d$ and plugging in information from the diagram, $C = 3.14 \cdot 18 \approx 56.52$ inches .

Guided Examples	Practice
Calculate the perimeter or circumference from the shapes given:	Calculate the perimeter or circumference from the shapes given:

Guided Examples:

1)

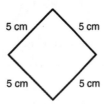

$Perimeter = 5+5+5+5 = 20$ cm

2) Diameter = 9cm

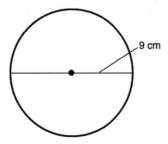

$C = \pi d \approx 3.14 \cdot 9 \approx 28.26 cm$

Practice:

1)

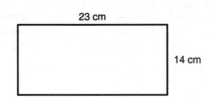

2)

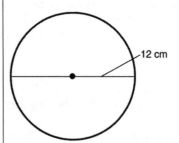

3) You and your family are moving to a new home. Of course, you have to have the biggest bedroom. You have not seen the home. Based on the blueprint measurements, which bedroom will you choose if bed room #1 has dimensions of 12,11,9,15 feet and bedroom #2 has dimensions of 12,12,10,10 feet ?

Objective 10.R.2 – Find the area of a polygon or circle

The area of a rectangle calculates the number of square units inside the shape. For comparison, think of perimeter as the length of fence needed to enclose your yard, where the area of the shape measures the area of the yard inside the fence.

With regards to units, perimeter is one dimensional (from Learning Objective 10.R.2) measured in *units* and area is measured in square units such as square inches, square feet, square centimeters or square meters and is written as *units*^{2} .

The formula for calculating the area of a rectangle or parallelogram is $A = b \cdot h$, where A is the area, b is the measure of the base of the shape and h is the height of the shape.

As we recall, the distance around a circle is called its circumference. The area of a circle measures the number of square units inside the circle.

The formula for calculating the area of a circle is $A = \pi r^2$, where A is the area, π is approximately 3.14 and r is the radius. The radius is measured as the distance from the center of the circle to any point on the circle.

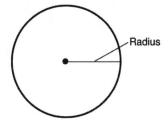

Radius

Guided Examples   Practice

Calculate the area of the following shapes:	Calculate the area of the following shapes:
1) 6 in. [rectangle] 35 in. The area of the rectangle is $A = b \cdot h = 35 \cdot 6 = 210$ inches^2	1) [rectangle] 21 cm 6 cm

2)

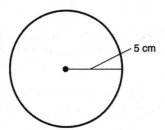

The area of the circle is $A = \pi r^2 = 3.14 \cdot 5^2 = 78.50$ cm^2

2)

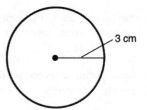

3) The student parking lot on your campus needs to be coated. The parking lot is a rectangle shape and measures 400 feet by 220 feet. How much parking lot surface (in square feet) needs to be covered with blacktop?

Name:_____Date:_____

Instructor:_____Section:_____

Objective 10.R.3 – Find the volume of the geometric solid

A quick examination recalls that perimeter focuses on one dimension measurement, area speaks to two dimensional measurement, and now, volume is the geometry of three dimensional measurement. This is due to an emphasis on three dimensions – width (w), length (l) and height (h), resulting in cubic units such as cubic inches, cubic feet, cubic centimeters and cubic meters.

In three dimensions, these geometric shapes consist of containers –boxes, food storage and cylinders. For us, our attention will emphasize volume. Specifically, how much material (water, sand, food) a container may hold. We concentrate on two specific shapes – the rectangular prism and the cylinder.

A rectangular prism comprises the same properties but takes different forms – from a shoebox to a cereal box to a pizza box and even a smartphone. The general form is:

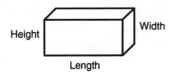

The volume formula for a rectangular prism is $V = l \cdot w \cdot h$ with $units^3$.

A cylinder comprises similar components – from a Pringle's can to canned drinks to some oatmeal containers. The emphasis is on the elements $\pi \approx 3.14$, radius (r) and height (h). The general form is:

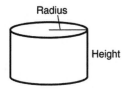

The volume formula for a cylinder is $V = \pi r^2 h$.

Guided Examples	Practice
Calculate the volume for the given geometric solids: 1) 18 in. 5 in. 15 in. For this rectangular prism, we use the formula $V = l \cdot w \cdot h$. So, $V = 15 \cdot 5 \cdot 8 = 600 inches^3$.	Calculate the volume for the given geometric solids: 1) 7 cm 5 cm 13 cm

2)

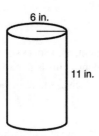

6 in.

11 in.

For this cylinder, we use the formula $V = \pi r^2 h$.
So, $V \approx 3.14 \cdot 6^2 \cdot 11 = 1243.44 inches^3$.

3) You are shopping for candles and notice two types of candles. The rectangular prism shaped candle measures 7 inches by 4 inches by 10 inches. The cylinder shaped candle has a radius of 3 inches and is 12 inches tall. Which candle shape has the greatest volume and the most wax?

Rectangular prism: $V = 7 \cdot 4 \cdot 10 = 280 inches^3$

Cylinder: $V \approx 3.14 \cdot 3^2 \cdot 12 = 339.12 inches^3$

The cylinder shaped candle has the greatest volume and contains more wax.

2)

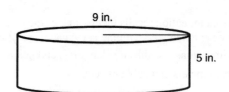

9 in.

5 in.

3) You are thirsty and are presented with two choices at the store. A juice box measures 5 cm by 4 cm by 6 cm. The same juice comes in a can that has a radius of 3 cm and is 6 inches tall. Which shape contains has the greatest volume and contains more juice?

Name:_____Date:_____

Instructor:_____Section:_____

Objective 10.R.4 – Find the missing sides of similar triangles

In this section, we explore and discover relationships between similar triangles. We first need to know what similar triangles are. Two triangles are similar if all their angles are the same measure and their corresponding sides are in the same ratio. This can be reworded to mean the same triangle shape but different size.

We need to recognize and identify the corresponding sides of similar triangles, and then write ratios to compare and identify any missing lengths of sides of similar triangles. Let's look at an example of the two similar triangles:

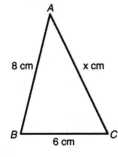 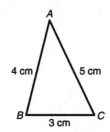

The example states these are similar triangles. As a general guideline, these triangles fit the similar triangle classification – the triangles have the same shape but a different size. Therefore, the corresponding sizes are similar.

We can identify the corresponding sides of these two similar triangles and express the relationship in a proportion or a set of three equal ratios $\dfrac{sideAB}{sideDE} = \dfrac{sideBC}{sideEF} = \dfrac{sideAC}{sideDF}$.

We apply algebraic reasoning to write ratios to form a proportion to solve for the length of the missing side. By substituting the given lengths, we generate $\dfrac{8}{4} = \dfrac{6}{3} = \dfrac{x}{5}$. Lastly, we use the cross product rule to solve for the missing value. We can use either complete proportion to solve for the missing value. To solve, $\dfrac{6}{3} = \dfrac{x}{5}$, then $6 \cdot 5 = 3 \cdot x$, and $30 = 3x$, then divide both sides by 3 to get $10 = x$. The missing length of $sideAC$ is 10 cm.

Find the missing side of the similar triangles:	Find the missing side of the similar triangles:
1) 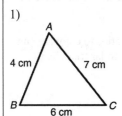 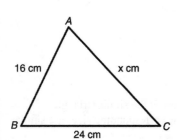	1) 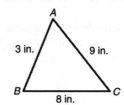 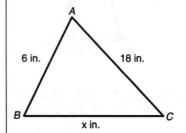

$\dfrac{sideAB}{sideDE} = \dfrac{sideBC}{sideEF} = \dfrac{sideAC}{sideDF}$, substitute $\dfrac{4}{16} = \dfrac{6}{24} = \dfrac{7}{x}$.

To solve, $\dfrac{4}{16} = \dfrac{7}{x}$, then $4x = 16 \cdot 7$, and $4x = 112$, then divide both sides by 4 to get $x = 28$. The missing length of *sideDF* is 28 cm.

Name:_____Date:_____

Instructor:_____Section:_____

Objective 10.R.5 – Find the unknown side of a right triangle, using the Pythagorean Theorem

Our challenge in this section is to comprehend, apply and calculate the dimensions of a right triangle using the Pythagorean Theorem. To use the Pythagorean Theorem accurately, we need to become familiar with the 90 degree angle (the right angle), the legs and the hypotenuse of the triangle.

With a right triangle, one of the angle measures exactly 90 degrees. The side that is opposite of the 90 degree angle is called the hypotenuse. This side is always the longest side of the right triangle. The other sides (always shorter than the hypotenuse) are called legs. The right triangle components look like:

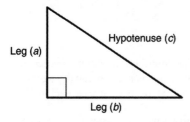

The Pythagorean Theorem states that if you add the squares of the lengths of the two legs of a right triangle, it equals the square of the length of the hypotenuse. Specifically, let a and b be the lengths of the two legs of a right triangle and let c be the length of the hypotenuse. Then,

$$a^2 + b^2 = c^2$$

Guided Examples   Practice

Use the Pythagorean Theorem to calculate the length of the hypotenuse:	Use the Pythagorean Theorem to calculate the length of the hypotenuse:
1) In a right triangle, side $a = 9$ inches and side $b = 12$ inches. What is the length of the hypotenuse? 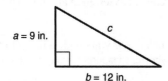 Since $a^2 + b^2 = c^2$, we substitute the values into the appropriate part of the equation. So, $9^2 + 12^2 = c^2$. To solve for c, $81 + 144 = c^2$, then $225 = c^2$. To solve for c, we need to take the square root of both sides. This looks like $\sqrt{225} = \sqrt{c^2}$. Finally, you may need a calculator to take the square root of 225 resulting in $15 = c$. By our calculation, the length of the hypotenuse is 15 inches.	1) In a right triangle, side $a = 18$ inches and side $b = 24$ inches. What is the length of the hypotenuse?

Chapter 11 Mathematics and the Arts

Learning Objectives

Objective 11.R.1 – Simplify radical expressions

Objective 11.R.1 – Simplify radical expressions

Music and Mathematics Review

The connections between music and mathematics have a rich and deep history. The contribution of mathematics to music focuses on tuning systems. One system is named the "even-tempered" scale which is mathematically starts with the understanding that an octave must have the ratio $2:1$. Since there are 12 half-steps in an octave, each half-step must have a ratio of $2^{\frac{1}{12}}$, described as 2 raised to the $\frac{1}{12}th$ power. Additionally, this can be considered as $\sqrt[12]{2}$ and written as the 12*th* root of 2. This mathematics concept shows up in the following partial frequency of notes table:

Note	Frequency (cps)	Ratio to frequency of preceding note
C	260	$\sqrt[12]{2} \approx 1.05946$
C#	275	$\sqrt[12]{2} \approx 1.05946$
D (second)	292	$\sqrt[12]{2} \approx 1.05946$
D#	309	$\sqrt[12]{2} \approx 1.05946$

Proportion and the Golden Ratio Review

The Golden Ratio is denoted by the Greek letter φ (phi) and represents a special number approximately equal to 1.618. This ratio is determined to be "most pleasing to the eye" and is frequently viewed in art and architecture and observed in nature.

The Golden Ratio is a mathematical relationship expressed as a fraction such that if you divide a line into two parts, the Golden Ratio mathematically exists if the ratio of the sum of the two shorter measurements to the larger measurement $\left(\dfrac{a+b}{a}\right)$, is equal to the ratio of the larger measurement to the smaller measurement $\left(\dfrac{a}{b}\right)$. Expressed algebraically, the Golden Ratio is, $\left(\dfrac{a+b}{a}\right) \approx \left(\dfrac{a}{b}\right) \approx 1.618$.

Visually, the Golden Ratio consists of the following:

This golden ratio is commonly used with the rectangle shape. We, as consumers, are bombarded with manufacturer attempts at the golden rectangle shape. Which rectangle is most pleasing to consumers – a debit/gift/credit card, a smartphone, a pop-tart or a Hershey's candy bar?

Now, if you look at a rectangle with the length greater than the width. This creates the proportion $\dfrac{A+B}{A} = \dfrac{A}{B} 1.618$. If the ratios are approximately equal to 1.618, then you have a golden rectangle.

For example, your credit card measures 8.5 mm (side A) in length and 5.5 mm (side B) in width. Does this shape approximate the golden rectangle, a shape that is "most pleasing to the eye"? Does the proportion approximate 1.618?

$\dfrac{A+B}{A} = \dfrac{A}{B} = \dfrac{8.5+5.5}{8.5} = \dfrac{8.5}{5.5} = 1.618$. So, $1.647 \neq 1.545 \neq 1.618$. Based on these calculations, a credit card is NOT in the shape of a golden rectangle.

Guided Examples   Practice

Use a calculator to approximate the following:	Use a calculator to approximate the following:
1) $\sqrt[12]{2} \approx$	1) $\sqrt[12]{2} \approx$
Use the Golden Ratio to determine which measurements would create a shape that is "most pleasing to the eye":	Use the Golden Ratio to determine which measurements would create a shape that is "most pleasing to the eye":

2a) $a = 6$ $b = 4$

Use the Golden Ratio format to check the relationship among the length of the sides $\dfrac{6+4}{6} \approx \dfrac{6}{4} \approx 1.618$.

Convert to decimal form and analyze the results.
$1.67 \neq 1.50 \neq 1.618$, so these measurements are not "most pleasing to the eye."

2b) $a = 8$ $b = 5$

Use the Golden Ratio format to check the relationship among the length of the sides $\dfrac{8+5}{8} \approx \dfrac{8}{5} \approx 1.618$.

Convert to decimal form and analyze the results.
$1.63 \approx 1.60 \approx 1.618$, so these measurements follow the Golden Ratio format. A shape that utilizes these measurements would be "most pleasing to the eye."

2a) $a = 15$ $b = 7$ OR

2b) $a = 13$ $b = 8$

3) A smartphone shape measures 13 mm in length and 6.5 mm in width. Does this shape create a golden rectangle with proportions that are "most pleasing to the eye"?

Chapter 12 Mathematics and Politics

Learning Objectives

Objective 12.R.1 – Evaluate an algebraic expression in table form

Objective 12.R.1 – Evaluate an algebraic expression in table form

In this chapter, you will discover that different methods of counting votes can lead to different results and further investigate the "fairest" method of apportioning Congressional representation among voters. To do this, we will practice evaluating an algebraic expression in table form. You will be given several values and a rule and then the value must be placed into the rule or process to determine the numerical outcome.

For example, fill in the table given the following information:

Value (x) is the number of hours you study	**The process $5x+60$ describes a relationship between the number of hours you study and the score on your next math test (out of 100).**
0	$5\cdot 0+60 = 60$
2	$5\cdot 2+60 = 70$
4	$5\cdot 4+60 = 80$
6	$5\cdot 6+60 = 90$

Guided Examples	Practice

Fill in the table given the following information:

1) The growth chart in a pediatrician's office may be modeled after the process formula $2x+7$, where x is the number of months. Fill in the table to calculate the weight (in pounds) of a newborn at various months of age.

Value in months	Process $2x+7$ (in pounds)
1	$2 \cdot 1 + 7 = 9$
3	$2 \cdot 3 + 7 = 13$
6	$2 \cdot 6 + 7 = 19$
9	$2 \cdot 9 + 7 = 25$

Fill in the table given the following information:

1) Since 2001, the process formula $0.4x+11$ describes the poverty rate in the United States, where x is the number of years after 2000. Fill in the table to calculate the poverty rate (in percent) for the given years:

Years since 2000	Process $0.4x+11$ (in percent)
1	
2	
3	
4	

Answers

Chapter 1

Objective 1.R.1 1) True 2) True 3) False 4) False

Objective 1.R.2 1) False 2) $6 + 4 \neq 10$ 3) True 4) False 5) True 6) True 7) False 8) True

Objective 1.R.3 1) rational, real 2) natural, whole, integer, rational, real 3) False 4) True

Chapter 2

Objective 2.R.1 1) $\dfrac{10}{21}$ 2) $\dfrac{6}{5}$ 3) $\dfrac{10}{7}$ 4) $\dfrac{20}{3}$

Objective 2.R.2 1) $3.28 \cdot 10^5$ 2) $1.4629 \cdot 10^4$ 3) $3.4 10^{-3}$ 4) $5.8 \bullet 10^{-6}$ 5) 6,820 6) 710,000 7) 0.000911
8) 0.02185

Chapter 3

Objective 3.R.1 1) $\dfrac{3}{11}$ 2) $\dfrac{1}{5}$ 3) $\dfrac{4}{14} = \dfrac{4 \div 2}{14 \div 2} = \dfrac{2}{7}$

Objective 3.R.2 1) $\dfrac{83}{100}$ 2) 0.14 3) 57% 4) 60% 5) $75 6) $90 7) 2 tons

Objective 3.R.3 1) 0.0738 2) 0.75 3) Yes 4) 25% 5) 5% 6) Yes

Objective 3.R.4 1) 3.15×10^5 2) 1.7×10^{13} 3) 3.7×10^{-3} 4) 1×10^{-6} 5) 2900 6) 149,600,000 7) 0.043
8) 0.000006 9) 8×10^8 10) 9.6×10^{10} 11) 2×10^9 12) 1.84×10^3

Objective 3.R.5 1) 430 2) 530,000 3) 9800 4) 8.214 5) 439.3 6) 45.34

Chapter 4

Objective 4.R.1 1) 100 2) 167 3) $260 4) $4,820 5) 18 6) 25 7) -11 8) -60 9) -10 10) 5 11)
10 12) 10 13) -6

Objective 4.R.2 1) $x = 16$ 2) $x = 9$ 3) $x = 6$ 4) 30 students

Objective 4.R.3 1) 2 2) 0.061 3) 0.08 4) 7.9% 5) 410% 6) 12.42 7) 285 8) $500 \cdot 4\% = \$20$

Objective 4.R.4 1) $x = 6$ 2) $x = 6$ 3) $x = 9$

Objective 4.R.5 1) 12 2) 4 3) 4.90 4) $1.29 = 129\%$

Chapter 5

Objective 5.R.1 1) Illinois 2) Kentucky 3) 9,900

Chapter 6

Objective 6.R.1   1a) 4 years old   1b) 4 years old   1c) 2 years old

Objective 6.R.2   1) 3   2) 8   3) approximately 6.93

Objective 6.R.3   1) $x = 60$ 2) $x = 30\%$ 3) $x = 125$

Objective 6.R.4   1) 0.058   2) 0.14   3) 0.071

Chapter 7

Objective 7.R.1   1) 72   2) 240 outfits   3) $\dfrac{25}{36}$ 4) $\dfrac{5 \cdot 4 \cdot 3}{10 \cdot 9 \cdot 8} = \dfrac{60}{720} = \dfrac{1}{12}$

Objective 7.R.2   1) 6.4   2) 11.6   3) $E(v) = \left(2 \cdot \dfrac{9}{10}\right) + \left(10 \cdot \dfrac{1}{10}\right) = 1.8 + 1.0 = \$2.80$

Objective 7.R.3   1) $(2)^4 \cdot (7)^3$ 2) $(9)^3 \cdot (4)^2$ 3)   24   4) 120   5) 114   6) 30   7) 10,000

Chapter 8

Objective 8.R.1   1) linear growth   2) exponential decay   3) linear decay   4) exponential growth

Objective 8.R.2   1) $x = \dfrac{1.602}{0.477} \approx 3.358$ 2) $x = \dfrac{1.875}{1.079} \approx 1.737$

Objective 8.R.3   1) $y = 5$ 2) $y = 3$ 3) $y = 3$

Chapter 9

Objective 9.R.1   Point A: $(3,1)$ Point B: $(-2,4)$ Point C: $(-3,2)$ Point D: $(1,-4)$

Objective 9.R.2   1) $y = \dfrac{2}{3}x - 5$

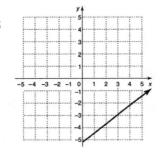

2) $y = \dfrac{1}{2}x + 0$

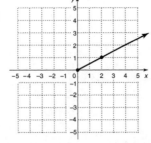

Objective 9.R.3   1) $10 = 10$ yes   2) $\$7,000 \neq \$7,520$ no

Objective 9.R.4   1) $slope = \dfrac{20-15}{8-6} = \dfrac{5}{2}$ 2) $slope = \dfrac{7-9}{8-5} = \dfrac{-2}{3}$

Objective 9.R.5   1) $x = \dfrac{\log 53}{\log 7} \approx \dfrac{1.72428}{0.84509} \approx 2.0404$ 2) $x = \dfrac{\log 220}{\log 6} \approx \dfrac{2.34242}{0.77815} \approx 3.0102$ 3) $x = 10^2$ 4) $x = 10^{3.6}$

Chapter 10

Objective 10.R.1   1) $Perimeter = 23 + 14 + 23 + 14 = 74cm$ 2) $C = \pi d \approx 3.14 \cdot 12 = 37.68cm$ 3) Bedroom #1

Objective 10.R.2   1) $A = 21 \cdot 6 = 126cm^2$ 2) $A = \pi r^2 \approx 3.14 \cdot 3^2 = 28.26cm^2$ 3) $88,000\,ft^2$

Objective 10.R.3   1) $V = 7 \cdot 5 \cdot 13 = 455cm^3$ 2) $V \approx 3.14 \cdot 9^2 \cdot 5 = 1271.70cm^3$

   3) Prism: $V \approx 5 \cdot 4 \cdot 6 = 120cm^3$, Cylinder: $V \approx 3.14 \cdot 3^2 \cdot 6 = 169.56cm^3$ CYLINDER

Objective 10.R.4 1) $x = 16$ in.

Objective 10.R.5 1) $c = 30$ inches

Chapter 11

Objective 11.R.1 1) $\sqrt[12]{2} = 1.05946$ 2) The choice 2b best fits the Golden Ratio because $\dfrac{13+8}{13} \approx \dfrac{13}{8} \approx 1.618$. In decimal form $1.61 \approx 1.62 \approx 1.618$.

Chapter 12

Objective 12.R.1 1)

Years since 2000	Process $0.4x + 11$ (in percent)
1	$0.4 \cdot 1 + 11 = 11.4\%$
2	$0.4 \cdot 2 + 11 = 11.8\%$
3	$0.4 \cdot 3 + 11 = 12.2\%$
4	$0.4 \cdot 4 + 11 = 12.6\%$